EMERGING TECHNOLOGIES
FOR THE CONTROL OF HAZARDOUS WASTES

Emerging Technologies
for the Control of
Hazardous Wastes

by

B.H. Edwards
J.N. Paullin
K. Coghlan-Jordan

Ebon Research Systems
Washington, DC

NOYES DATA CORPORATION
Park Ridge, New Jersey, U.S.A.
1983

Published in the United States of America by
Noyes Data Corporation
Mill Road, Park Ridge, New Jersey 07656

10 9 8 7 6 5 4 3 2 1

Library of Congress Cataloging in Publication Data

Edwards, B. H.
    Emerging technologies for the control of hazardous
wastes.

    (Pollution technology review, ISSN 0090-516X ;
no. 99)
    Bibliography: p.
    Includes index.
    1. Hazardous wastes. I. Paullin, J. N.
II. Coghlan–Jordan, K. III. Title. IV. Series.
TD811.5.E38 1983   628.5'4   83-4022
ISBN 0-8155-0943-X

Foreword

This book reviews and assesses emerging technologies or novel variations of established technologies for the control of hazardous wastes. Most of the hazardous wastes considered in the study are organic substances. Mining, plating, metallic, and nuclear wastes are not included. Information on the various technologies was gathered by an extensive literature survey.

Current interest in hazardous waste handling methods and disposal practices is apparent almost daily in the news media. Methods which might reduce the ultimate disposal problems facing industry and/or decrease the actual quantity of wastes generated are actively being sought.

Three major technologies are covered in detail in the book—molten salt combustion, fluidized bed incineration, and ultraviolet (UV)/ozone destruction. Theory, unit operations, specific wastes treated, and economics are discussed for each of these.

Several other technologies, which are in the development stage, are also described. In this category are catalyzed wet oxidation, dehalogenation by treatment with UV irradiation and hydrogen, electron bombardment of trace toxic organic compounds, UV/chlorinolysis of aqueous organics, and catalytic hydrogenation-dechlorination of polychlorinated biphenyls (PCBs).

Among the wastes to be treated by these emerging technologies are various dioxins, PCBs, pesticides, herbicides, chemical warfare agents, explosives, propellants, nitrobenzene, plus hydrazine and its derivatives.

The information in the book is from *Emerging Technologies for the Control of Hazardous Wastes* (EPA Report 600/2-82-011), prepared by B.H. Edwards, J.N. Paullin, and K. Coghlan-Jordan of Ebon Research Systems for the U.S. Environmental Protection Agency, September 1981.

The table of contents is organized in such a way as to serve as a subject index and provides easy access to the information contained in the book. A complete list of references is included.

Advanced composition and production methods developed by Noyes Data are employed to bring this durably bound book to you in a minimum of time. Special techniques are used to close the gap between "manuscript" and "completed book." In order to keep the price of the book to a reasonable level, it has been partially reproduced by photo-offset directly from the original report and the cost saving passed on to the reader. Due to this method of publishing, certain portions of the book may be less legible than desired.

Acknowledgment

Ebon Research Systems would like to thank our Project Officer, Mr. Thomas Baugh for his support, assistance, and guidance throughout this project.

Notice

This report has been reviewed by the Municipal Environmental Research Laboratory, U.S. Environmental Protection Agency, and approved for publication. Approval does not signify that the contents necessarily reflect the views and policies of the U.S. Environmental Protection Agency or the Publisher, nor does mention of trade names or commercial products constitute endorsement or recommendation for use.

Contents and Subject Index

-1-
Introduction

PURPOSE OF THE REPORT

The purpose of this report is to identify and evaluate new and emerging technologies for hazardous waste treatment and disposal. The major technologies evaluated are:
- molten salt combustion
- fluidized bed combustion
- ultraviolet/ozone destruction

In addition to these technologies, processes employing catalyzed wet oxidation, dehalogenation by the addition of hydrogen in the presence of ultraviolet (UV) light, high energy electrons, UV/chlorinolysis and catalytic hydrogenation-dechlorination are now in the developmental stage.

During the course of the study, major hazardous waste generators were surveyed to determine whether they would be interested in information on these and other techniques for hazardous waste disposal. The results of this survey are also presented.

SOURCE MATERIAL FOR THE REPORT

The material for the identification and evaluation of these technologies has been gathered by an intensive literature survey conducted over the course of a year. Although extensive use has been made of manual and computerized data bases, it was also necessary to monitor the recent literature and upcoming conferences and symposia to gain access to material not yet published. Personal communications were also used in the survey.

CHARACTERIZATION OF HAZARDOUS WASTE

Hazardous wastes in this study are defined and characterized by the Hazardous Waste Proposed Guidelines and Regulations and Proposal on Identification and Listing published in the Federal Register on December 18, 1978.

Most hazardous wastes in this study are organic substances. A listing of hazardous wastes combusted by various technologies is included in the appendicies. Mining wastes, plating, and metallic wastes are not considered in this report. Nuclear wastes and wastes from health care facilities are beyond the domain of this report.

Conclusions

MOLTEN SALT COMBUSTION

Molten salt technology has been in existence for many years, but only recently has molten salt combustion been used for the treatment of hazardous wastes. In the process, hazardous material is combusted at temperatures below its normal ignition point, either beneath or on the surface of a pool of molten salt. Individual alkali carbonate salts, such as sodium carbonate, or mixtures of these salts, are usually used as the melt, but other salts can be employed based on waste characteristics. Containers for the molten salts are made of ceramics, alumina, stainless steel, or iron.

Ideally, during the molten salt combustion process, organic substances are totally oxidized to carbon dioxide and water, while heteroatoms such as phosphorus, sulfur, chlorine, arsenic, and silicon are reacted with the carbonate melt to form $NaCl$, Na_3PO_4, Na_2SO_4, Na_2SO_5, and Na_2SiO_3. Iron from metal containers forms iron oxide. Most organic substances are destroyed, leaving behind a relatively innocuous residue, while harmless levels of off-gases are emitted. Generally, the salt bath is stable, nonvolatile, nontoxic, and may be recycled for further use until the bath is no longer viable.

Some hydrocarbons combusted by the molten salt process are chlorinated hydrocarbons, PCB's, explosives and propellants, chemical warfare agents, rubber wastes, textile wastes, tannery wastes, various amines, contaminated ion exchange resins, tributyl phosphate, and nitroethane.

The technology has progressed from bench-scale through the pilot plant stage to the construction of a demonstration-size coal gasification unit. Additionally, portable units mounted on truck beds have been used.

Advantages of Molten Salt Combustion of Hazardous Waste

Some of the advantages of molten salt combustion are:
- Combustion is nearly complete
- Non-polluting off-gases are emitted
- Operating temperatures are lower than in normal incineration; thus they are fuel efficient.
- The system is amenable to recycling generated heat.
- The system does not require highly skilled operators, i.e. a professional engineer's license is not required.

- A wide variety of hazardous wastes can be combusted.
- Bulky wastes can be combusted after pre-sizing.
- Many wastes can be combusted in compliance with EPA regulations.

Major Problems with Molten Salt Combustion of Hazardous Waste

Some problems with molten salt combustion of hazardous wastes are:
- Particulate emissions from some wastes are high, although generally less than from normal incineration.
- The technology is not readily adaptable to aqueous wastes.
- The molten salt bath must be bubbling (but not ebullient) to promote efficient combustion.
- Eventually waste salt and ash must be disposed of, or the fluidity of the melt will be destroyed.
- A hazardous waste with greater than 20% ash cannot be combusted.
- Detailed economic information for a demonstration-sized system is not currently available (1980).

FLUIDIZED BED INCINERATION

Fluidized bed systems have had many industrial uses since the technology was proposed by C. E. Robinson about a century ago, yet fluidized bed combustion of hazardous waste is a relatively new technique. A hot fluidized bed offers an ideal environment for combustion. Air passage through the bed produces strong agitation of the bed particles. This promotes rapid and relatively uniform mixing of carbonaceous materials. Bed mass is large in relation to the quantity of injected waste, and bed temperatures, which usually range from 750-1000°C, are quite uniform.

Hazardous wastes that have been incinerated in a fluidized bed include chlorinated hydrocarbons with a high chlorine content, waste PVC, waste PVC with coal, PVC insulated waste wire, munitions (TNT, RDX, and Composition B), spent HCl pickling liquor, spent organotin blasting abrasive, and a waste organic dye-water slurry. A listing of various materials used as the bed medium is included in the Unit Operations Section.

Advantages of Fluidized Bed Incineration of Hazardous Wastes

Advantages of fluidized bed combustion of hazardous waste are:
- The combustor design concept is simple and does not require moving parts after the initial feed of fuel and waste.
- Fluidized bed combustion has a high combustion efficiency.
- Designs are compact due to high volumetric heating rates.
- Nitric oxide formation is minimized because of low gas temperatures and low excess air requirements. Low excess air requirements also reduce the size and cost of gas handling equipment.
- In some cases, the bed itself can act to neutralize some of the hazardous products of combustion.
- The bed mass provides a large surface area for reaction.
- Temperatures throughout the bed are relatively uniform.

- Continued bed agitation by fluidizing air allows larger waste particles to remain suspended until combustion is completed.
- If the hazardous waste contains a sufficient calorific value, a fluidized bed combustor can operate without auxiliary fuel.
- Excess heat generated by fluidized bed combustion of wastes with high caloric value can be recycled.
- Fluidized beds are able to process aqueous waste slurries.
- As the bed functions as an efficient heat sink, major variations in feed consistency and water content take a long time to effect temperature changes in the bed medium.
- The heat sink effect also limits radiation from the bed and allows the combustion system to be shut down for considerable periods of time (weekends) and restarted with little or no pre-heat time.

Major Problems With Fluidized Bed Incineration of Hazardous Waste

Some problems with fluidized bed combustion of hazardous wastes are:
- Bed diameters and height are limited with design technology; therefore, maximum volumetric flow rates per unit are limited.
- Removal of inert residual material from the bed (such as ash) can be difficult in some instances.
- In systems where temperature must be controlled at lower limits because of other thermal considerations, increased residence time can cause carbon buildup in the bed.
- Certain organic wastes will cause the bed to agglomerate, thereby reducing its effectiveness.
- Particulate emissions are a major problem with fluidized bed combustion. In some cases, particulates are high even when emissions are passed through a Venturi scrubber.

UV/OZONE DESTRUCTION

Ozone treatment is an established technology for the treatment of some hazardous wastes. Recent studies show that a combination of ultraviolet light with the ozonation process is a more effective technique for destroying hazardous waste than the use of ozonation alone.

The addition of UV light to the ozonation process has greatly expanded the number of compounds that can be destroyed. Exposed halogen atoms, unsaturated resonant carbon ring structures, readily accessible multi-bonded carbon atoms, and alcohol and ether linkages are particularly susceptible to UV/ozone systems. Compounds with shielded multi-bonded carbon atoms, sulfur compounds, and phosphorous compounds are less susceptible to UV/ozonation.

A combination of ozone with UV treatment has been used to destroy PCB's, 2,3,7,8-tetrachlorodibenzo-p-dioxin (TCDD), nitrobenzene and related derivatives, and the hydrazine family of fuels.

Many complex decisions representing trade-offs are necessary to implement a well-designed, efficiently operated, economic UV/ozonation system. This technology has advanced considerably in recent years, and is close to state-of-the-art for many hazardous wastes. Some wastes which are difficult to oxidize are not treatable by this technology. Reaction kinetics plays a large role in the decision to apply UV/ozone technology to the treatment of a specific waste. Although UV/Ozonation is much more limited in the variety and concentration of wastes that can be treated (when compared to conventional incineration, molten salt combustion, or fluidized bed combustion), it is still a viable technology for the treatment of certain hazardous wastes.

Advantages of UV/Ozonation

Some advantages of UV/ozonation are:
- Aqueous or gaseous waste streams can be treated.
- Capital and operating costs are not excessive as compared to incineration
- Chemical carcinogens and mutagens can be treated.
- The system is readily adaptable to on-site treatment of the hazardous waste.
- UV/ozonation can be used as a final treatment to supplement partial treatment systems.
- It can be used as a preliminary treatment for certain wastes.
- It can be used to meet effluent discharge standards.
- Modern systems are usually automated, thereby reducing labor.

Disadvantages of UV/ozonation

Some disadvantages of UV/ozonation are:
- Ozone is a non-selective oxidant; therefore, the waste stream should contain primarily the hazardous waste of interest.
- UV/ozone systems are generally restricted to 1% or lower levels of hazardous compounds. Most hazardous substances that have been treated by this process were in the ppm levels.
- Ozone decomposes rapidly with increasing temperature; therefore, excess heat must be rapidly removed.

Superiority of Molten Salt Combustion, Fluidized Bed Incineration, and UV/ozonation to Landfills

All of the emerging technologies—molten salt, fluidized bed, and UV/ozonation—can be considered as alternatives to landfill disposal of hazardous waste. The intent of the emerging technologies is waste destruction, or at least attenuation, to acceptable levels. Landfills either store waste in specialized containers or attempt to prevent its spread from the area where it was dumped.

The future fate of hazardous wastes stored in landfills is, in many cases, unknown. Deepwell injection, encapsulation, and other forms of containment that do not attempt to destroy the hazardous waste share the same uncertain future. If cost is not considered, the use of technologies that destroy hazardous wastes should be considered far superior to landfill storage. However, insufficient information exists to compare the emerging technologies on a cost/benefit basis with established landfill practices.

Molten Salt Incineration and Fluidized Bed Incineration vs. Conventional Incineration

Both these technologies employ sufficiently high temperatures to efficiently destroy many types of hazardous wastes. Molten salt technology can meet EPA regulations regarding a minimum high temperature and residence time for the treatment of certain wastes, even though it employs much lower temperatures than are used in conventional incineration. Compliance with the regulations is achieved with much higher combustion efficiency, reducing the emissions to lower levels. EPA regulations regarding combustion criteria (Federal Register, Volume 43, Number 243, pp. 59008, 59009, December 18, 1978, #250.45-1 Incineration) state that the incinerator shall operate at temperatures greater than 1000°C, greater than 2 seconds residence time, and greater than 2% excess oxygen for the incineration of hazardous wastes. Halogenated aromatic hydrocarbons, such as PCB's, must be incinerated at more than 1200°C, and greater than 2 seconds residence time, with greater than 3% excess oxygen. Combustion efficiency must equal or exceed 99.9%. However, there is an exception. Other conditions of temperature, residence time, and combustion efficiency are permitted if an equivalent amount of combustion is demonstrated. Because of its higher combustion efficiency, molten salt combustion complies with the regulation exception. PCB's are combusted by molten salt technology at lower temperatures, yet the combustion efficiency requirement of 99.9% was exceeded by molten salt technology with a combustion efficiency greater than 99.9999% and a nominal residence time of 0.25-0.50 sec. Although both processes have some problems with particulate emissions, they have less problems with particulates than conventional incineration techniques.

Molten salt combustion and fluidized bed incineration can be considered as capital intensive for start-up costs. The same can be said for conventional incineration. Little information exists for the costs of molten salt combustion. The cost statistics for fluidized bed incineration are better known at the pilot plant stage. Fluidized beds and molten salt baths function as heat sinks, and can, in some instances, use the hazardous waste combusted as a fuel. Thus, they are potentially less expensive than conventional incineration techniques.

HAZARDOUS WASTE GENERATOR SURVEY

The results of the survey of hazardous waste generators indicated that there is not a strong interest in some of the new, emerging hazardous waste destruction technologies. This is probably because many companies are not familiar with the advantages of the emerging methods compared to the more established technologies such as landfills and conventional incineration. However, a high percentage of the respondents were interested in an information service such as a computerized data base or a newsletter regarding hazardous waste information.

Ebon Research Systems believes that an information gap exists for the dissemination of information regarding new hazardous waste technology. A computerized data base and/or newsletter should consider the following information:

- How to use established techniques more efficiently and economically—this would include information such as Btu values of specific combustion products.
- National, state, and county hazardous waste legislation—including packing and transportation regulations.
- Spill cleanup techniques
- Waste exchange and recycling
- Emerging destruction technology information. Types of wastes treated, economics, and feed mechanisms would be detailed.

-3-
Theoretical Basis of the Technologies

MOLTEN SALT COMBUSTION

The Catalytic Properties and Physical Structure of Molten Salts

When heated to a temperature slightly above their melting point, salt mixtures exist in a transition state between solid and liquid. Similar transition states exist at even higher temperatures between liquid and gas. This transition state is analogous to an equilibrium of ionic and non-ionic conditions. Measurements of conductivity sharply decrease in the solid state. (This effect is often used to good advantage in an electrically heated salt bath where the medium is also its conductor). The ionic activities give rise to the reactive nature of the salt bath and are proportional to the electronegativity of the ionic species present. The higher the stability (i.e. melting point) of a salt medium, the greater its reactivity upon thermal ionization.

According to Greenberg, most non-charged materials are soluble in molten salts. This solubility is related to the crystal structure of salts in the molten state. Data from x-rays taken at temperatures above the melting points of salts indicate that molten salts retain a quasi-lattice structure. It is assumed that the solute assumes an electronic charge in the semi-crystalline melt. This charge gives the solute an electrostatic orientation similar to that of the ionic component of the molten salt and enchances the solubility of the solute (1,2).

Various pollutant species, especially hydrocarbon derivatives, react with oxygen at relatively high energies of activation. However, when these compounds are exposed to oxygen in molten salt baths, they are oxidized at temperatures substantially lower than those normally necessary. Theoretically, the new orientation of the formerly neutral species results in a reduction of the energy required to initiate and sustain chemical reactions. The ability of molten salts to dissolve various neutral substances and act as catalysts for their oxidation is the basis for the feasibility of using molten salts in hazardous waste destruction (1,2).

The salts used in the molten salts process should be stable at temperatures required for combustion of specific hazardous substances. A single salt or mixture of salts may be used. The higher the stability of a salt medium (i.e. melting point), the greater its reactivity upon thermal ionization. Eutectic mixtures are often useful as they provide good efficiency of operation at lower temperatures. In a eutectic mixture, the salts are combined in a ratio that results in making the melting point of the mixture less than the melting point of any of its components. In eutectic mixtures, the melting range is evidence of the multiple equilibrium existing between the two salt componnents (2,3).

There appears to be no one all-purpose melt for molten salt combustion. The choice of salt(s) is dictated by the dual demands of reactivity and melting range. One of the more important theoretical considerations is the number of conversions that occur within the melt, as conversions result in greater reactivity, promote efficiency, and reduce unwanted emissions. However, the types and combinations of salts that can be employed in this process allow it to operate at a wide range of temperatures and under varied conditions of oxygen availability.

Specific Melts Used in the Process

Anti-Pollution Systems Neutral and Active Salts
  and Their Eutectic Mixtures--

Neutral salts—Molten salts can be assigned to one of two major classes. The first class can be considered as neutral salts. These salts, or mixtures of salts, lower the energy of activation for oxidation of the solute yet, they do not add oxygen or react chemically with it. They require an external oxygen source supplied by combustion gas, ambient air, or added oxygen (1).

Metallic halides with melting points in the 50°-600°C range are useful in molten salt baths as neutral salts. The chlorides are especially stable. A representative, but not all inclusive, list of neutral salt eutectic mixtures which may be used in molten salt baths appears in Appendix A. The listed temperatures represent the melting point of the bath and are accurate to + 10°C. Any neutral salt with a melting point less than 600°C can be used by itself, i.e., LiBr (547°C), LiI (446°C), and CaI (575°C) (1).

Active salts—Chemically oxidizing, active salts can also be used in molten salt combustion to enhance the equilibrium pressure of oxygen both at the surface and within the melt. These salts not only catalytically induce oxidation, but also maintain an equilibrium oxygen gas pressure that facilitates oxidation by continuously donating nascent oxygen and retaking ambient oxygen.

A partial list of chemically active oxidizers is shown in Appendix B. The temperature listed is accurate to ± 10°C and represents the lowest temperature at which the melt may be used. Some of these mixtures are combinations of neutral and active salts. These baths, like the neutral melts, all possess sharp melting points.

It is possible to lower the melting point of most baths by addition of lithium salts. The use of sulfate anion rather than carbonate is recommended. Lithium sulfate forms a stable monohydrate.

Oxidizing salt mixtures produce baths with lower melting temperatures than neutral salt mixtures. Mixtures often improve the normal, oxygen-releasing characteristics of the salts. For example, in the mixture of nitrate and nitrite, nitrite has a tendency to cause nitrate to more readily release oxygen. While neutral salt baths are normally used at temperatures immediately above their melting points, the oxidizing salt baths are normally used at temperatures approximately 93.3°C above their melting point in order to facilitate oxygen release. The lower end of the bath temperature range is limited only by the melting point of the salt (or mixture of salts). A temperature as low as 50°C can be achieved by selecting materials such as a mixture of thallium nitrate (50 M%) and silver nitrate (50 M%). The composition and operating temperature of a bath is largely determined by the composition and heat content of the material to be oxidized (1,2).

Although oxidizing baths are in a neutral or inert state at temperatures just above melting to 93.3°C above their melting point, they still retain their catalytic capacity and can be used within the inert range in order to avoid extreme temperature rises which may occur when highly flammable material is added to the melt. Because molten neutral salt baths operate at generally higher temperatures, the use of an inert oxidizing melt permits operation at temperatures where exothermic reactions are in much less danger of producing undesirable explosions (1,2).

Molten Salt Technology at Rockwell International

In an early study done at Rockwell International (1969), Heredy et al. removed sulfur oxides from flue gas using a mixture containing Li_2CO_3, Na_2CO_3, and K_2CO_3 (400°C melting point). This mixture reacted with sulfur dioxide and carbon dioxide to form alkali metal sulfite. The molten product of the reaction, consisting primarily of alkali metal sulfite dissolved in molten alkali metal carbonate, was treated with a gaseous mixture of hydrogen and carbon monoxide at a temperature between 400-650°C. The mixed alkali carbonates were regenerated, and at the same time, hydrogen sulfide, a marketable product, was formed (7). This process illustrates that a chemical reaction between solvent and solute can be effective in the destruction of waste. A mixture of 45+M% Li_2CO_3, 30+ M% Na_2CO_3, and 25+ M% K_2CO_3 was used for convenience. This was not the true eutectic (5).

Recent (1979) combustion studies have used a Na_2CO_3 melt. The chemical reactions of the waste with Na_2CO_3 depend on the chemical composition of the waste. Carbon and hydrogen molecules are converted to CO_2 and H_2O (steam). Halogens form corresponding sodium halide salts. Phosphorus, sulfur arsenic, and silicon form Na_3PO_4, Na_2SO_4, Na_2AsO_5, and Na_2SiO_3. Any iron present (from containers) forms iron oxide. Small quantities of nitrogen oxides are formed by fixation of nitrogen in the air. Modifications of this basic melt will be discussed in sections on destruction of specific wastes (6,7,8).

FLUIDIZED BED INCINERATION

Concept of the Fluidized Bed

If a fluid (gas or liquid) flows upward through a bed of solid particles, it exerts a frictional drag on the particles. This creates a corresponding pressure drop across the bed. So long as the force is smaller than the weight of the bed, as in fixed bed combustion, the particles will remain essentially motionless, and the fluid will flow through the interstitial passages (9).

As the fluid velocity is raised, a point is reached where the drag force just equals the bed weight. This is the point of incipient fluidization. The fluid velocity at this point is called the minimum fluidization velocity. As fluid velocity is increased above incipient fluidization, the drag force is sufficient to support the weight of the particles, the particles are bouyed up, exhibit great mobility, and behave like a fluid. The bed is then fluidized and flows under a hydrostatic head (10,11).

A fluidized bed resembles a liquid in several ways (10).
- The upper surface stays horizontal when the container is tilted.
- When 2 beds are connected, their levels equalize.
- Solids will gush in a jet from a hole bored in the side of the vessel, and fine solids can be made to flow much like a liquid through ducts and pipes.
- Although the upper surface of the bed is in motion, it is well defined.
- A light object can float on the surface of the bed; a heavy object will sink.
- The pressure drop between any two points is esentially equal to the static head of the bed between two points.

The relatively uniform temperature within a fluidized bed is due to several factors. Turbulent agitation within the fluidized mass breaks and disperses any hot or cold spots throughout the bed before they can grow to significant size. There is also a rapid movement of solids from one part of the bed to another. This does not mean that every solid particle in a fluidized bed is the same temperature. However, departure of an individual particle from the mean temperature of a fluidized bed will be much less than in a fixed bed (11).

Continued bed agitation by fluidizing air allows larger waste particles to remain suspended until combustion is completed. Bed depths usually range from about 40 cm to several meters. Variation in bed depth can affect the residence time of combustibles. It is desirable to minimize bed depth consistent with complete combustion and minimum excess air. In general, a shallow fluid bed depth is preferred in a continuous process because this provides the lowest pressure drop and power consumption as well as maximum heath and mass transfer.

Turbulent agitation within the fluidized mass breaks and disperses any hot or cold spots throughout the bed before they can grow to significant size. There is also a rapid movement of solids from one part of the bed to another. This does not mean that every solid particle in a single fluidized bed maintains the same temperature. However, the departure of an individual particle from the mean temperature of a fluidized bed will be much less than in a fixed bed (11).

Continued bed agitation by fluidizing air allows larger waste particles to remain suspended until combustion is completed. Bed depths usually range from about 40 cm to several meters. Variation in bed depth can affect the residence time of combustibles. It is desirable to minimize bed depth consistent with complete combustion and minimum excess air. In general, a shallow fluid bed depth is preferred in a continuous process because this provides the lowest pressure drop and power consumption as well as maximum heat and mass transfer. If longer residence time is required a high length-to-width ratio in bed size can optimize residence time distribution (11).

The Character of Fluidization

The character of fluidization depends on whether the bed is fluidized by a liquid or gas. In liquid fluidization, the neighboring particles typically move further apart to accomodate increases in flow rate above minimum fluidization, and the bed appears to undergo a stable uniform expansion. This state is called particulate fluidization.

If gas is the fluidizing medium, a bed of fine particles (50-200 μ) also exhibits particulate fluidization just above minimum fluidization. When gas passes through the interstices of the dense phase solids, excellent gas-solids contact is provided. When the gas velocity is increased, a second threshold, the minimum bubbling velocity, limits expansion. Any additional gas traverses the bed in the form of voids or bubbles which are virtually devoid of particles This is called aggregative fluidization. The gas bubbles provide poor gas-solids contact. The mean bed particle size and size distribution determine the quantity of incipient fluidization gas for fixed feed conditions. In general, the smaller the bed particle size, the larger is the quantity of gas that is able to pass through in the form of bubbles (11).

Solid activity and the rate of particle carry-over increase with gas velocity, yet a dense bed is retained over a wide range of gas velocities. At some sufficiently high velocity, massive entrainment occurs, the bed's upper surface is destroyed, and the bed may no longer be regarded as stationary. In the absence of a cyclone, the bed would soon empty. Most commercial fluidized bed combustors operate at gas velocities well below this point (10). Studies on hazardous waste destruction in fluidized beds that are discussed in this report use aggregatively fluidized, bubbling beds.

Bubbles in Fluidized Beds

The structure and behavior of fluidized beds are influenced by bubbles. If a fluidized bed is viewed through transparent walls, the bed appears like a boiling liquid. The upper surface of the bed in in motion, but it is well defined. Bubbles rise through a fluidized bed and set it into continuous vigorous agitation. They then burst at the surface, flinging particles into the space above. Some bubbles, especially the smaller ones, are elutriated from the bed. The small particles which escape are normally trapped by cyclones and returned to the bed. Internal or external cyclones may be installed (10).

If the bed is high and narrow, bubbles tend to coalesce and fill the entire cross section of the bed. These large bubbles are called slugs, and the bed is said to be a slugging bed. Data taken under slugging conditions can be misleading for designing large diameter reactors (10,11).

In a bed of large particles, bubbling begins at incipient fluidization. The bubbles grow rapidly because of the density difference between the bubble and the surrounding bed. This causes a pressure gradient that drives gas into the bubbles. Also, because of the high viscosity of the large-particle bed, bubbles rise slowly in them. Accordingly, gas in the vicinity of the bubble uses it as a short cut, entering through the bottom and exiting at the top.

If a bed is made of small particles, the rate of growth of bubbles is small because of the high resistance to gas flow in the bubble. Since bed viscosity is low, bubbles rise rapidly through the bed. The slow rate of bubble growth and their short residence time in the bed decrease the tendency for slug formation. As in the bubbles in large-particle beds, gas also enters at the bottom of the bubble. It is then swept around and dragged down the sides of the rising bubbles. The region around the bubble is then called the cloud. Bed particles, outside of the cloud, do not contact the gas. This is why poor conversions have been observed in bubbling, fine-particle beds (10).

Fast Fluidized Beds

It is possible to operate at a gas velocity sufficient to blow most of the solid material out of the reactor in a relatively short time if fresh solid bed material is simultaneously added. This fast fluidized bed operates at gas velocities above the bubbling regime and is free of large voids or bubbles. There is some opinion that fast fluidized beds offer several advantages over the bubbling bed. Yerushalmi and Cankurt (11) claim that fast beds have higher processing capacities, more efficient and intimate contact between gas and solid, and better capability in handling cohesive solids. There is no indication that fast fluidized bed combustion technology has been used in destruction of hazardous waste.

UV/OZONE DESTRUCTION

Ozone generally is regarded as a resonance hybrid of four contributing structures:

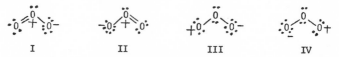

I   II   III   IV

Ozone reacts as a 1,3-dipole or as an electrophilic reagent through the electron deficient terminal oxygen atoms (structures III and IV). It also reacts through the negatively charged terminal atom on all four structures (12,13).

Ozonation of dilute solutions of organic species produces organic oxidation and hydroxyl radical formation. Bridging of carbon-carbon double bonds by ozone forms unstable ozonide intermediates that decompose into smaller oxidation species. This process continues until carbon dioxide and water or relatively stable refractory compounds (such as acetic or oxalic acids) are formed (13).

Qualitatively, the overall reaction is depicted on the three regions shown on the ozonolysis curve in Figure 1.

The mechanisms for many of these reactions are complex and, in some cases not well understood. Readers interested in a more thorough treatment of these reaction mechanisms are referred to surveys by Oehlshlaeger (12) and Bailey (13).

In many experimental situations, reaction conditions do not produce complete oxidation, and a number of intermediates remain in the reaction solution. It is possible that many of these intermediates are as harmful as, or more harmful than, the parent compound. It has recently been found that, if ozone treatment is combined with UV (ultraviolet radiation), the UV radiation activated the refractory compound and all subsequent refractory species. This permits nearly complete oxidation to carbon dioxide, water, and other elementary species (14).

The following groups are especially vulnerable to attack by ozone enhance UV absorption:
1. exposed halogen atoms
2. unsaturated resonant carbon ring structures
3. readily accessible multi-bonded carbon atoms
4. alcohol and ether linkages
Shielded multi-bonded carbon atoms, sulfur, and phosphorus are much less vulnerable (14).

There is a substantial difference in the chemistry involved and results achieved when the UV/ozonation process for treating hazardous wastes is compared to straight ozonation.

When ultraviolet radiation, in the 180-400 nm range, is added to the process, 72-155 kcals/mole of additional energy is provided, and the O_3 molecule breaks down into oxidizing O species. This is ample energy for producing substantially more free radicals and excited-state species for . the initial compound and subsequent oxidation species than those produced by ozonation alone. As a result, excited atomic species (O), hydroxy (OH) and hydroperoxy (HO_2) radicals, and excited-state species (M)* are produced from reactant molecules. These all greatly enhance the overall oxidation rate, prevent plateauing, and permit complete oxidation. The following is a simplified representation of how these excited species are probably formed:

$$O_3 + h\nu \longrightarrow O_2^* + O$$

$$O + H_2O \longrightarrow 2OH$$

$$O_3 + OH \longrightarrow HO_2 + O_2$$

$$M + h\nu \longrightarrow M^* \longrightarrow R, I + H$$

Overall:

$$\left.\begin{array}{l} M \\ R \\ I \end{array}\right\} + OH, HO_2, O, O^* \longrightarrow CO_2, H_2O, Cl^-, SO_4^{-2}, PO_4^{-3}$$

where: M is the pollutant species being oxidized, R is a free radical species, and I is an intermediate molecular species (14).

According to Prengle, elevated UV light also accelerates ozone decomposition so that using UV excessively results in less favorable economics (14). However, no generalization of requirements should be made because of wide variation in reactivities of species. Free cyanide is oxidized by ozone without elevated temperature or UV, but iron complexed cyanides have maximum practical temperature and UV requirements. It is better to treat acetic acid at naturally occurring pH rather than neutralize, yet phenol destruction is not strongly pH dependent. practical temperature and UV requirements. It is better to treat acetic acid at naturally occurring pH rather than neutralize, yet phenol destruction is not strongly pH dependent (15).

In summary, the overall mechanism of ultraviolet/ozone photooxidation of M-species in aqueous solution occurs by a combination of the following:
- O3 photolysis to produce oxidizing O species
- H_2O photolysis and reaction with O_3 to produce OH and HO_2 oxidizing species
- M photolysis to produce M and free radicals

The free radicals mentioned above participate in a sequence of oxidation reactions leading to the final oxidation products. For M-species that are low UV absorbers, the rate controlling mechanism is photolysis oxygen species oxidation, but for high UV absorbers, M-species photolysis is rate controlling.

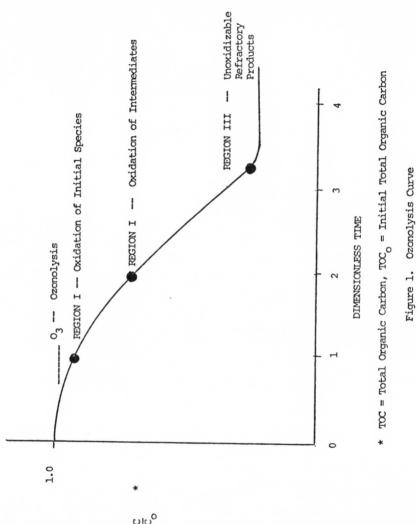

Figure 1. Ozonolysis Curve

-4-
Unit Operations

MOLTEN SALT COMBUSTION

The Rockwell International Combustors

In the Rockwell International process, combustible material and an oxygen source, usually air, are continuously introduced beneath the surface of a molten salt (7). The type of molten salt used is usually determined by the nature of the waste to be incinerated. Sodium carbonate or potassium carbonate, alone or in combination, is frequently used as the melt. Sodium sulfate (approximately 10%) is sometimes added to the melt to increase oxygen availability. Melts are usually maintained at 800-1000°C (6,7,8).

The method of waste addition is designed to force any gas formed during combustion to pass through the melt before it is emitted into the atmosphere. The system is engineered to render any gaseous emission into relatively innocuous substances. Theoretically, the intimate contact of the waste, melt, and air causes a high heat transfer to the waste and results in its rapid and complete destruction. As previously indicated, the chemical reactions of the waste with molten salt and air depend on the waste composition. Ideally, the off-gas should contain carbon dioxide, steam, nitrogen, and unreacted oxygen. If particulates of inorganic salts are present in the off-gas, they are removed by a Venturi scrubber or by passing through a baghouse (6).

Ash concentrations above 20% must be removed to ensure fluidity of the melt. Batchwise melt removal is sufficient for low throughput applications. When the throughput is 250 kg/hr or higher, a side stream of the melt is continuously processed. During the continuous side stream removal, care must be taken to ensure sufficient salt remains in the melt. Residual material, such a ash, is removed from the spent melt by quenching in water followed by filtration. Recovered salt is recycled to the combustor (6,7).

Most of the published recent work (1977-1980) on molten salt combustion of hazardous waste has been performed by Rockwell International in their four molten salt combustion facilities. Two are bench-scale combustors used for feasibility and optimization tests with feed rates of 0.25-1 kg/hr of waste. The third faciltiy is a pilot plant combustor with a feed rate of 25-100 kg/hr of waste. It is used to obtain engineering data for design and reliability extrapolation to a full sized plant. The fourth system is a portable unit, designed for the disposal of empty pesticide containers.

The Department of Energy-funded coal gasification process development unit, with a design through-put of 1,000 kg/hr of coal, is also a Rockwell International combustor (6,7,8).

Bench-scale Combustion System—

Unit design—A schematic of the bench-scale molten salt combustion system is shown in Figure 2. The combustor contains about 5.5 kg of molten salt in a 15-cm ID, 90-cm high alumina tube placed in a Type 321 stainless steel vessel. The stainless steel vessel is in turn contained in a 20-cm ID, four heating zone Marshall (Ohio) furnace. The temperatures of the four heating zones of 20-cm height are controlled by silicon-controlled rectifiers. Furnace and reactor temperatures are recorded by a chart recorder. An air cooling system, which prevents a temperature increase in the system when an excessive amount of heat is released into the melt, consists of an eight-hole air distribution ring mounted under the stainless steel retainer vessel. Air, at rates near 500 l/min can be passed upward between the outer surface of the retainer vessel and the furnace wall. A 3.7-cm ID alumina feed tube is adjusted so that its tip is immersed approximately 1-cm above the bottom of the 15-cm diameter alumina reactor tube. Thus, the waste-air mixture is forced in a downward path through the feed tube, outward at the bottom and, finally, circulates upward through about 14 cm of molten salt. This insures complete and rapid mixture of the waste with the melt (7).

One of the bench-scale combustors has been modified for the incineration of very hazardous wastes. The combustor is located in a walk-in hood, and there is controlled access to the room which contains the hood. In order to increase personnel safety, all process controls are located outside the hood. Gas from the room is scrubbed in an activated charcoal adsorber (7).

Waste feeding—The bench-scale molten salt combustion unit can accommodate solid, liquid, and mixtures of liquid and solid wastes. After solids are pulverized by a No. 4 Wiley Mill, they are placed in the hopper and metered by screw feeding into the 1.2-cm OD central tube of the injector. The screw feeder is rotated by a 0-400 rpm Eberback Corporation Con-Torque stirrer motor. Hangups in the hopper are prevented or released by a Syntron Model V-24 vibrator. The solids are mixed with about 75% of the air used for combustion. The other 25% of the combustion air passes through a cooling annulus in the injector (not shown in Figure 2). The air-solids stream combines with the cooling air stream at the tip of the injector. The mixture emerges into the alumina feed tube and then enters the melt (7).

Liquid waste is pumped into the injector with a Fluid Metering, Inc. (FMI) laboratory pump. As the liquid travels down through the central tube of the injector, the combustion air passes down through the cooling annulus. The liquid waste and air streams combine at the injector tip and pass downward through the alumina feed tube into the melt (7).

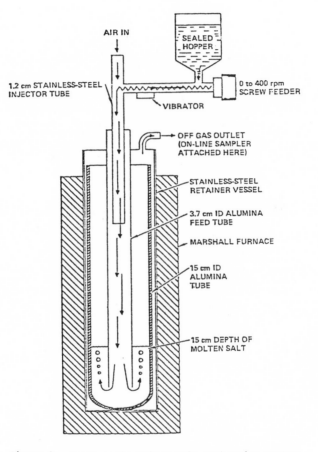

Figure 2.   Bench-scale Molten Salt Combustion System

Because the FMI pump is suitable only for solid-free liquids, slurries are fed into the system by a Masterflex 7014 or 7016 peristaltic pump. A 5-liter stainless steel beaker located on a digital balance contains the slurry and a stirrer. The pump and feed beaker are connected by flexible tubing. Since the digital balance is capable of reading 10 kg by 1-gm increments, continuous mass feed can be calculated.

For types of waste where the solid is soaked with liquid and does not have a suitable consistency for feeding as a dry solid or pumping as a liquid, activated carbon or some other suitable absorbing agent is added to produce a nearly-dry, free-flowing material (7).

Off-gas analysis for bench-scale combustors—A typical on-line combustion off-gas analysis schematic is shown in Figure 3. A sampling of the off-gas is first passed through a high efficiency particulate air (HEPA) filter to remove particulates. The particulate-free gas is then analyzed by Rockwell International for NO_x, CO, O_2, N_2, and unburned hydrocarbons. A water-ice trap is used to remove most of the water vapor upstream of all the analyzers except the NO_x analyzer. Because some NO_x might condense, an ice trap is not used upstream in the NO_x line. Rockwell International analyzes NO_x with a Thermo Electron Corporation Chemiluminescent NO_x Analyzer. CO_2 analyses are made with an Olson-Horiba, Inc. Mexa-200 Analy Analyzer. The CO and hydrocarbon analyses are made with an Olson-Horiba, Inc. Mexa-300 Analyzer. Determinations for O_2 are made with a Teledyne portable oxygen analyzer, Model 320.

For continuous gas determinations and experimental control, the Olson-Horiba and Teledyne instruments are used. A Perkin-Elmer gas chromatograph is used to confirm these analyses. Syringe samples (1 ml) of the off-gas are removed downstream from the ice trap and injected into the gas chromatograph. Chromatographic columns Porapak Q and Molecular Sieve 5A are operated at 100°C. A typical Rockwell International gas chromatographic analysis assays CO_2, CO, N_2, $O_2(+Ar)$, and unburned hydrocarbons (7).

Pilot Plant Combustion System—

A schematic of a Rockwell International molten salt pilot plant is shown in Figure 4. The molten salt vessel is made of Type 304 stainless steel and lined with 15-cm thick refractory blocks. The 4-m high, 0.85 m ID vessel contains 1000 kg of salt. Melt depth is 1 m when air does not circulate through the bed. A natural-gas-fired burner is used to preheat the vessel on start-up and maintain heat during standby. The heat content of the waste is usually sufficient to maintain the melt in a molten condition during combustion periods (7,8).

The molten salt vessel is loaded via the carbonate feeder. Combustible materials to be processed are crushed to the required size in a hammermill, transferred into a feed hopper equipped with a variable-speed auger, and finally, introduced into the air stream for transport into the vessel (7).

Exhaust gases generated in the vessel pass through refractory-lined tubes in the vesel head before they enter a refractory-line mist separator.

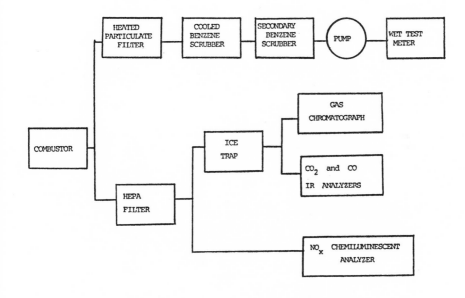

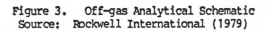

Figure 3.   Off-gas Analytical Schematic
Source: Rockwell International (1979)

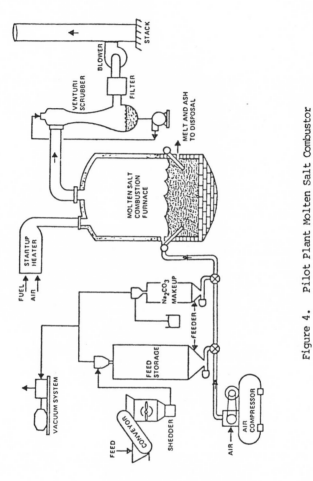

Figure 4.   Pilot Plant Molten Salt Combustor

Source: Rockwell International (1979)

The mist separator traps entrained melt droplets on a baffle assembly. The gases are then ducted into a high-energy Venturi scrubber or baghouse to remove particulate matter prior to release into the atmosphere. During extended testing, an overflow weir can be used to permit continuous removal of spent salt (8).

Portable Molten Salt Disposal System—

In the Rockwell International portable molten salt unit for disposal of empty pesticide containers, the combustor and auxiliary components are mounted on a truck bed. The combustor, which is 1.8-m ID and 3.4-m tall, can process 225 kg/hr of waste. The empty containers are conveyed to a shredder, shredded (particle size 0.95-3.2 cm), and pneumatically conveyed to the combustor. Off-gas is cleaned by a particle separator to remove smaller particulates before the gas is emitted from the stack (17).

Theoretically, such a unit would yield only about 250 liters of spent salt for each 907 kg of containers processed. Spent salt is drained into a drain cart and buried in a Class 1 dump. After cooling, the system is transported via truck to a new site where fresh salt is added to the combustor and the cycle repeated (17,18).

Rockwell International Coal Gasification System—

The Rockwell International process demonstration unit (PDU) is another example of the scale-up potential for a molten salt system. This process development unit, funded by a Department of Energy contract, is able to gasify coal into low, intermediate, or high Btu gas. The system is designed to accept all coals, including highly-caking, high sulfur bituminous coals. Crushed coal is gasified at high temperature and high pressure by reaction with air in a highly turbulent mixture of molten sodium carbonate containing sodium sulfide, ash, and unreacted carbonate. The process demonstration unit can accept 907 kg/hr (19).

The Anti-Pollution Systems Combustion Unit

Anti-Pollution Systems (APS) has developed an alternative molten salt process based on the molten salt technology discussed earlier (1,2,4). In one application of the process, textile manufacturing waste containing acrylics residue was purified. The contaminated liquid waste, gravity-fed at a rate of about 1,995 l/hr was introduced onto the surface of a molten salt bath composed of 62 mole% KNO_3-38 mole% $Ca(NO_3)_2$. Although the bath melts at 140°C, it was maintained at 450°C. Under these conditions, water in the waste was completely flash-evaporated, leaving behind an organic residue which ignited and rapidly decomposed under the bath's catalytic influence (2).

The bath was housed in a container constructed of noncorrosive material. The container had a main chamber where evaporation and combustion of the liquid takes place. In an adjacent baffled chamber, the gases of combustion produced in the main chamber are again brought into contact with the melt to further combust any incompletely oxidized products. Waste is introduced into the chamber from above. The contact surface of the bath is about 1.5 m x 6 m with a 5.1 cm–deep melt. Heat is supplied by an open gas flame. Air, provided at a flow rate of about 1400-2800 l/min, is maintained at a steady flow by a blower at the input and a pump at the output. Air, carbon dioxide, and water vapor are off–gases. No measurable hydrocarbons or carbon monoxide were detected in the off–gases (4).

The APS molten salt unit has been modified in several ways so that it can function not only as an incinerator for carbonaceous materials but also purify automotive exhaust systems and some types of stack emissions (1,2). Because the APS molten salt scrubber is capable of removing particulates in the submicron range, Greenberg suggests that molten salt scrubbing systems could be effective for the destruction of toxic substances such as Kepone (4).

In a modification in which the salt bath operates as an afterburner (shown in Figure 5), the system consists of a stainless steel box within a box. The inner box (trough) floats on 7.6 cm of salt. The floating inner trough makes ash removal simple and is designed to preclude problems when water is introduced directly into the melt. If alkali sulfate is used in the melt, it is estimated to have a heat capacity of 593°C–equivalent to a 38.1 cm bed of cast iron. Liquids (containing water at any concentration) or solids are introduced into the center box and exposed to a flame (4). As the waste combusts, the generated heat is transferred to the salt bed beneath the combustion chamber. If the heat of combustion is sufficiently large, and enough waste is burned, it is not necessary to premelt the salt bed. The generated heat maintains the bottom portion of the combustion chamber at the melting point of the salt. Exhaust gases produced by combustion are pulled through a series of baffles and bubbled through the melt before exiting. This provides a second incineration for toxic, volatile substances and traps particulates in the melt (4,18).

Suction is supplied by an induced draft fan which creates a negative pressure on the baffle closest to the exit side of the system. This causes a rise in the liquid level with an accompanying drop in the melt level on the exit side of the baffle. The exhaust gases impinge on the liquid front created by the suction. A modified dry cyclone collector prevents salt losses at air velocities above 123,000 l/min. Carbon particulates and inert materials are removed by a fine mesh stainless steel screen (4,18).

Although the temperature of the APS melt is usually maintained between 560-600°C (lower than the 800-1000°C generally used in the Rockwell International unit) with an estimated residence time of approximately 0.1 second in the melt, the combustor can also be operated under residence times and temperatures which satisfy EPA regulations for chlorinated organics (4).

There are currently (1979) three molten salt units at APS. One is portable and can be fueled with propane (Personal communication from J. Greenberg to Dr. Barbara Edwards of Ebon Research Systems, July 12, 1979). Analysis methods, detailed off–gas descriptions, and precise operating parameters are not currently (1979) available in the literature for the APS molten salt process.

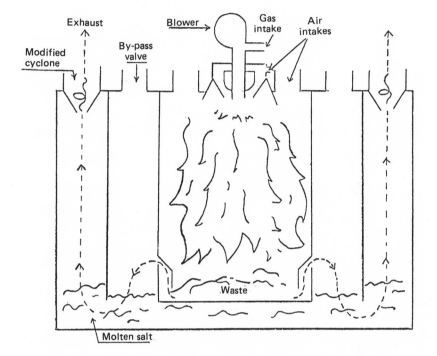

Figure 5. APS Molten Salt Incinerator
Source: Wilkinson et al. (1978)

Disposal of Waste Melts

Ash and other non-combustibles, such as glass or iron oxides, build up during combustion. When ash concentration exceeds 20 wt%, the ash must be removed to preserve melt fluidity. Additionally, the reaction of inorganic compounds such as halogens, phosphorous, sulfur, and arsenic with sodium carbonate ultimately causes sufficient melt conversion that results in loss of pollutant removal capability (6).

At Rockwell International, melt is transferred to a dissolving tank and treated with water or aqueous sodium bicarbonate solution. The dissolved carbonate solution is filtered to remove insoluble materials and treated with carbon dioxide gas to precipitate sodium bicarbonate. The bicarbonate salt is recovered by filtration and used in the molten salt furnace where it is converted to carbonate. Eventually, the bicarbonate solution contains excess chloride, phosphate, and sulfate and must be discarded (18).

When waste throughput exceeds 250 kg/hr, with sufficient sodium carbonate remaining in the melt, a melt side stream is continuously withdrawn and processed to recover and recycle the carbonate. Smaller waste throughput is removed batchwise. These small quantities of unprocessed melt and spent melt which no longer can be recycled are drained into a drain cart and buried in a Class 1 dump (a non-sanitary landfill underlain by unusable groundwater) (6).

In a patent for the disposal of explosives and propellants, Yosim either regenerated the melt or reacted it with molten lime or aqueous lime to form a water-insoluble calcium salt residue. These insoluble salts were formed by the reaction of lime with carbonate, fluoride, phosphate, and sulfate components in the spent melt (20).

Yosim also treated pesticide disposal melts with air to oxidize any residual sulfide to sulfate. The melt was then cooled or treated with water and lime for disposal in an approved dump site. Yosim suggested that a melt containing only chlorides from chlorinated pesticides decomposition could be dumped in the ocean without pretreatment (21).

Greenburg, in a patent for the oxidation of carbonaceous materials with molten salts, removed floating residues from the surface of the bath and used a dredging apparatus to remove settled products (1).

Although no details for the disposal of spent melt were cited by Edgewood Arsenal scientists when they combusted various military chemical agents, a disposal problem existed with the Na_3AsO_4 containing spent melt from the combustion of arsanilic acid (23).

It is possible that this type of hazardous spent melt can be insolubilized in glass or concrete. Salt samples from spent Rockwell International non-hazardous melts were glassified using different fluxing materials as the solidifying agents. The Na_2SO_4 and NaCl content of the resultant solidified glass varied from 10-40 wt%. Glassification temperatures ranged from 1100°-1500°C. An accelerated leach test, using continually flowing distilled water at 100°C, produced accelerated leach rates of 10^{-4} to 10^{-6} g/cm/day compared to 10^{-7} to 10^{-9} gm/cm/day for normal leach tests. The authors suggested clay, Pyrex, and probably basalt can be used to adequately glassify waste molten salts from the combustion or organics, using present molten salt technology (22).

Insolubilization in concrete was tested by Rockwell International scientists on waste salts containing Na_2CO_3, Na_2SO_4, and NaCl mixed with cement. The salt content of the salt-cement mixture was 10-38 wt%. The authors stated that since the low level leach rates of these concrete samples were only about 5-10 times more than glassified waste salt, the less expensive cement formation would be an acceptable disposal method for waste melts (17).

FLUIDIZED BED INCINERATOR

A schematic of a typical fluidized bed incinerator is shown in Figure 6. Air is driven by a blower and passes through a distributor plate into the bed above the plate. Sand is usually, but not always, used in the bed. Typical materials used as a fluidized bed are listed below:

- sand
- alumina
- sodium carbonate
- dolomite
- ferrous oxide (granular)
- the waste itself, e.g., spent organotin blasting abrasive

The upward flow of air (or some other fluidizing substance) through the bed results in a dense, turbulent mass. Material to be incinerated is injected into the combustor by specialized pumps, screw feeders, or pneumatically. Air passage through the bed promotes rapid and uniform mixing of the injected material within the bed. A cold reactant will attain the temperature of the bed almost instantly. Water quickly evaporates as combustion takes place. Auxiliary fuel (oil or gas), if needed, is usually introduced directly into the bed. Suspended fine particles are collected in a cyclone. Steam and the other gaseous products of combustion exit, along with suspended fine particles, from the top of the reactor. Exhaust gases are cleaned in a scrubber and exit to the atmosphere (24). As the specific combustion parameters vary with the type of waste combusted, more specific details on unit operations are discussed in the section on the types of wastes destroyed by the emerging technologies.

ULTRAVIOLET/OZONE DESTRUCTION

Ozone (O_3) is a three atom allotrope of oxygen. It is second only to fluorine in electronegative oxidation potential and has long been recognized as a powerful disinfectant and oxidant of both organic and inorganic substances. Generated by solar energy, ozone is a natural ingredient of the earth's atmosphere. It is also generated from atmospheric oxygen by energy from lightning and is associated with operation of most electrical equipment. Ozone is a gas under ambient conditions. Unreacted ozone decomposes in a matter of hours to simple, molecular oxygen. Ozone is not a poisonous chemical in the sense that it enters into internal body

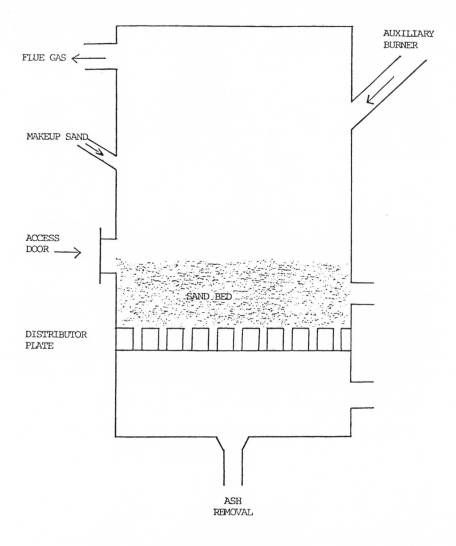

Figure 6.   Schematic of a Fluidized Bed Combustor
Source: Powers (1976)

chemistry. However, because of its strong oxidizing properties, exposure to low concentrations of ozone is damaging to delicate nasal, bronchial, and pulmonary membranes (26).

Ozone Generators

It has never been conclusively proven that ozone can be formed by purely chemical means. Unlike most chemicals, there is no controlled available natural source for ozone, and it is not practical to store it in containers. Ozone is generated when an oxygen molecule is sufficiently excited to disassociate into atomic oxygen; further collisions with oxygen molecules then cause the formation of the ozone molecule. Excitation energy can be supplied by ultraviolet light or high voltage (25,26).

A major source for ozone is an ozone generator. Most ozone generators use high voltage, although ultraviolet ozone generators are practical for outputs less than 1 gm/h. High voltage generators produce a corona discharge (also known as a silent arc discharge or a brush discharge). When electrons flow at sufficiently high potential through a gas that contains oxygen, a bluish glow accompanies the excited molecular state (25,27).

The following conditions are necessary to create a corona discharge:
- two electrodes separated by a gap
- a gas in the gap
- sufficient voltage potential between the two electrodes to cause current flow through the dielectric and gas

The electrodes can be flat, tubular, or any configuration that allows their opposing surfaces to be parallel. The distance between the parallel surfaces of the electrodes should be large enough to insure uniform current flow. Depending on the system, a gap smaller than a certain critical size may restrict the air flow to a point of excessive pressure drop through the unit. A large gap increases voltage requirements (25,26).

The two basic types of commercial ozone generators in use are the concentric tube generator and the parallel plate generator. The concentric tube system, first devised by Siemens, can be modified to operate at higher pressures than the plate design (26). One electrode is a metal grid or coating inside a cylindrical glass tube. A metal cylindrical tube fits over the glass tube and functions as the second electrode. The glass acts as a dielectric to increase the voltage gap between electrodes without sparking. Ozone is formed in the oxygen or air flowing longitudinally in the gap between the tubes. Manifolds distribute the fresh gas flow to multiple tubes and collect ozone—containing gas at the exits (28).

In one design, shown in Figure 7, there is a stainless steel center electrode through which water passes for heat removal. Concentric to that, on the outside, is a glass dielectric tube. The outside surface of the glass is coated with a metal that serves as the second electrode. The dielectric material is used to prevent arcing from one electrode to another. The second coated electrode and the glass electrode are oil—cooled (non-conductive fluid) which in turn is water—cooled in a heat exchanger (29).

In the more complex Otto system, high voltage, parallel plates alternate with ones that are water cooled. Gas flows in the air gaps between
the plates. There is more generating surface within a given volume than
in the concentric tube system. This design is being replaced by the
air-cooled Lowther system. Water cooling is more costly than air cooling,
but many years of experience with water systems has developed relatively
trouble-free techniques (26).

In 1979, only one manufacturer produced ozone generators with parallel
plates. The electrodes are mounted on an aluminum frame for dimensional
stability and arranged as small, square modules. Dielectric coatings are
applied to the electrodes to increase the voltage gradient, and ozone is
formed in the air gap between the plates. Manifolds are used to distribute
and collect the gas. Major U.S. manufactures of ozone generating equipment
are listed in Table 2 (28).

Parameters for Ozone Generation

Dielectric—

Operational efficiency of an ozone generator is dependent on parameters
involved in design and engineering (26). The dielectric material ideally
should have both a high electrical resistance and high thermal conductivity
Since these properties rarely occur together, the dielectric is usually
chosen for its high electrical resistance and depth to a minimum thickness
to overcome heat transfer deficiency. Glass or other ceramic materials are
usually used. Some polymers have demonstrated superior properties for
short time periods. Because both ozone production and ozone concentration
greatly depend on the quality of the dielectric materials used on the
electrode, U.S. Ozonair Corp. suggests these minimum requirements (26).

- Dielectric constant $\epsilon = 85$
- Electric Strength   15 Kv/mm
- Volumetric resistivity   10 ohm/cm

Pressure—
The operating pressure should provide sufficient force to deliver
ozonized gas to the contacting vessel. The pressure may have to overcome
a back pressure of from 0.6-0.9 meters of water at the reactor. Because
pressure affects the electrical impedance of the system, the operating
pressure must be correlated with the air gap and preferred voltage (25).
Flow Rate—
The time that feed gas is in the electrical discharge determines the
concentration of ozone in the effluent. This concentration can vary from
0.5-10 wt% in a well designed generator operating at ideal conditions. The
power efficiency drops rapidly as the ozone concentration increases (25).
Temperature—
Approximately 90% of the energy applied to the ozonator is lost as
heat. The ozone decomposition rate increases with increasing temperature,
and provisions should be made for the rapid removal of excess heat. Not
only does ozone output vary as a function of generator temperature, but
also, as ozone generator temperature increases, the dielectric material
changes thermal characteristics and is subject to rupture. As critical

generator dielectric temperatures are relatively low (120°C-130°C), the economics of regrigeration or coolant quantity supplied to the system must be evaluated versus the temperture of the ozone stream (9,10). Water is usually considered to be a more efficient medium for heat transfer than air. Air used for cooling ozonators can be of any quality, and when it passes through a generator, it is not degraded or changed. Typical generator cooling fluid requirements are 2831 m^3 of air/454 gm O_3 or 1789 liters of water/454 gm O_3 (30).

Feed Gas—

Ozone generation using oxygen is approximately twice as efficient as when air is the feed gas (i.e., ozone generated from air requires at least twice the power of an equal amount of ozone generated from pure oxygen). However, oxygen cost is more than twice the cost of air. Unless oxygen is recovered via a closed system, it is cheaper, especially when ozone generated is less than 225 kg/day, to generate ozone from air. A recently developed technique uses a "swing cycle" that concentrates oxygen from air by alternate passage through molecular sieves (25,30).

The corona intensity of new ozone generators requires perfectly clean, dry, and oil-free feed gas. If not, either ozone production is reduced, or the electrodes are damaged and need frequent cleaning. A minimum air dew point of −45°C is recommended to insure dependable ozone production (29).

If ozone generators operate at high pressures (as in the tubular design), dehumidification is enhanced and micron and submicron filters can be used in the air-feed system.

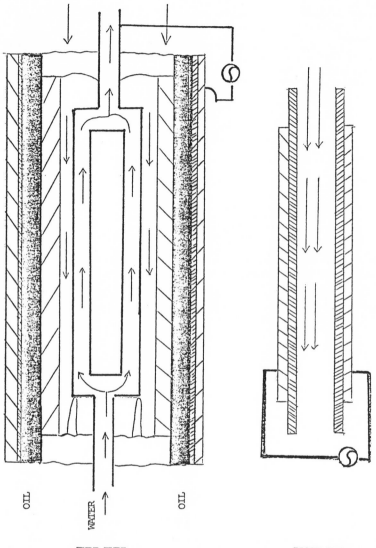

TUBE TYPE   PLATE TYPE

Figure 7. Ozone Generators

TABLE 1

MAJOR U.S. MANUFACTURERS OF OZONE GENERATING EQUIPMENT

Manufacturer & Address	Equipment	Models-Capacities lb O_3/day air feed	Cooling Method	Typical O_3 Concentration in air, %
Crane Cochrane P.O.Box 191 King of Prussia,Pa 19406 215-265-5050	Concentric tubes SS/glass/aluminum Series C-cabinet Series P-skid mounted	Series C, 1-18 lb/day Series P, 18-122 lb/day	Water on outer electrode	1
Emery Industries,Inc. Ozone Technology Group 4900 Estee Ave. Cincinnati, Oh. 45232 513-482-2100	Concentric tubes SS/glass/nichrome Skid mounted	Series 9270, 1-23 lb/day Series 9260, 21-400 lb/day	Water on outer electrode	1
Ozone Research & Equip.Corp. 3840 N. 40th Ave. Phoenix,Az. 85019 602-272-2681	Concentric tubes SS/glass/SS Series V,B,& D-cabinet Series H-skid mounted	Series B & V,1/4-2 lb/day Series D & H, 4-250 lb/day	Water on outer electrode	1
PCI Ozone Corporation One Fairfield Crescent West Caldwell, NJ 07006 201-575-7052	Concentric tubes SS/glass/silver Series G-cabinet Series B-skid mounted	Series G, 1-28 lb/day Series B, 35-1400 lb/day	Water on inner electrode, Oil on outer electrode	2

TABLE 1 (Continued)

MAJOR U.S. MANUFACTURERS OF OZONE GENERATING EQUIPMENT

Manufacturer & Address	Equipment	Models-Capacities lb O_3/day air feed	Cooling Method	Typical O_3 Concentration in air, %
Welsbach Ozone Systems Corp. 3340 Stokely St. Philadelphia, Pa. 19129 215-226-6900	Concentric tubes SS/glass/SS Series CLP & GLP both skid mounted	Series CLP,24-127 lb/day Series GLP,170-322 lb/day	Water on outer electrode	1
Infilco Degremont, Inc. Koger Executive Center Box K-7 Richmond,Va 23288 804-285-9961	Concentric tubes SS/glass/aluminum Skid mounted	No model designations, 10-600 lb/day	Water on outer electrode	1
Union Carbide,Linde Div. Environmental Systems P.O.Box 44 Tonawanda, NY 14150 716-877-1600	Parallel ceramic coated steel Lowther plates	No model designations, 1-1200 lb/day	Air on outside both electrodes	1
U.S. Ozonair Corp. 464 Cabot Rd. S.San Francisco,Ca 94080 415-952-1420	Concentric tubes Titanium/ceramic/aluminum	Series HF,5-570 lb/day	Water on inner electrode, Air on outer electrode	2

-5-
Wastes Destroyed by Emerging Technologies

HAZARDOUS WASTES DESTROYED BY THE MOLTEN SALT PROCESS

Background

In 1975, the U.S. Environmental Protection Agency sponsored a study by Battelle Pacific Northwest Laboratories to assess molten salt technology for the pyrolysis of solid waste. Although no actual studies were performed on hazardous wastes, Battelle recommended that molten salt technology receive further investigation for processing materials such as waste pesticides and herbicides with low ash content, waste nerve gases, biological warfare agents, and noxious fumes (31). A discussion of hazardous wastes treated by the molten salt process follows.

Explosives and Propellants

Obsolete and old explosives and propellants are usually destroyed by burning in an open area or detonation in a safety zone. These destruction processes emit pollutants such as smoke, hydrogen chloride, nitrogen oxides, carbon monoxide, and undesirable dust clouds. A patented molten salt process was developed by Rockwell International scientists for the disposal of explosives and propellants with minimal resultant environmental pollution (20). The process was developed primarily from combustion tests performed on 5 g samples of Composition B (60 wt% RDX, also known as Cyclonite, and 40 wt% TNT) and Standard Solid Propellant (70 wt% ammonium perchlorate, 10 wt% Aluminum, 14 wt% polybutadiene, plus other unidentified chemical substances).

The molten salt bath was located in the center of an armor-plated, three-sided cubicle with viewing windows. Five gram pieces of explosive were rolled through a non-metalic inclined pipe which led from the outside of the cubicle to the molten salt bath. The reaction between the melt and the explosive plus its effect on the surrounding atmosphere was observed through the windows. The bath was contained in a stainless steel vessel surrounded by an insulated clamshell heater. Melt temperature was monitored prior to the introduction of the waste.

Runs were performed with a ternary lithium carbonate–sodium carbonate–potassium carbonate eutectic and a NaOH-KOH eutectic at 400-600°C. A control run burned the explosive in the open without any melt present. There was vigorous bubbling in the melt before the explosive ignited, during the burn, and after combustion ceased. Times for complete combustion ranged from 3-11 minutes (20).

According to Yosim, a wide variety of commercially significant explosives and propellants may be effectively treated by the molten salt process. The particle size and weight of the explosive will, to some extent, determine the feed rate of the material to the molten salt combustor. The chemical composition determines the type of melt used. Temperature control may be required if a given substance has a low explosion temperature. Yosim states that the reactivity of the alkaline molten salts used and the physical state of the explosive or propellant (solid, liquid, mixture) are probably not significant factors in the effectiveness of the process. Moreover, it is also probably not important whether the explosive is present as a single or binary composition, or whether the propellants are double-base compositions containing various additives or polymeric matrix propellants. Thus, any of the bipropellant systems containing liquid fuel, such as the hydrocarbon fuels of the JP types, ethyl alcohol, hydrazine and hydrazine derivatives are feasible for molten salt decomposition.

The following explosives are considered by Yosim as suitable for the molten salt destruction process (20):

ammonium nitrate	diethylene glycol dinitrate
glyceral nitrate	TNT
diglyceral tetranitrate	Tetryl
glycol dinitrate	Cyclonite
trimethylolethane trinitrate	HMX
PETN	Composition B
	DEPHN

Chemical Warfare Agents

A series of molten salt combustion tests on various chemical warfare agents were performed at Edgewood Arsenal by their personnel using a bench-scale combustor built and designed by Rockwell International staff. Agents tested included GB spray-dried salts (the product of the chemical neutralization of GB), GB, VX, distilled mustard (HD), Lewisite (L), and obsolete war gas identification sets. The agents present in the sets along with the packaging material and dunnage in the sets were incinerated together in order to evaluate the feasibility of molten salt incineration for their disposal. Although the combustor was similar to the Rockwell International bench-scale apparatus described earlier, there were some modifications in the unit and analytical methods employed (23).

Combustion of GB Spray-Dried Salts—

GB ($C_4H_{10}PF$) was neutralized by aqueous sodium hydroxide to produce a solution containing mainly sodium isopropylmethylphosphonate (NaGB), NaF, NaOH, Na_2CO_3, and varying amounts of disodium methylphosphonic acid and diisopropyl-methylphosphonate. Spray-drying removes water from these compounds and produces "GB spray-dried salts."

Under certain conditions, NaGB, diisopropylmethylphosphonate, and disodium methylphosphonic anhydride can react with NAF to reform GB. As the removal of as much as 99.95% of the fluoride ion is insufficient to prevent reformation of trace quantities of GB, the complete destruction of the organophosphorous residue is necessary to eliminate the hazard of GB production during handling or storage of the spray-dried salts.

Although incineration can combust organophosphorous material, the noxious by-product, P_2O_5, is produced. Because there is no conventional, environmentally safe method for the disposal of GB spray-dried salts, the molten salt process, with its ability to scrub undesirable substances in the melt during the reaction, was tested. Earlier molten salt studies by Rockwell International scientists on Malathion revealed P_2O_5 emissions were substantially reduced, but not eliminated. A bench-scale program was therefore established for the molten salt incineration of GB spray-dried salts to determine the efficiency of the combustor for converting organophosphorous chemical agents in Na_3PO_4 without excess P_2O_5 emissions. The parameters varied were melt temperature (900°-950°C), melt depth (21 cm and 31 cm), and melt composition. The melt composition was either 90 wt% Na_2CO_3 with 10 wt% Na_2SO_4, or 100 wt% Na_2CO_3. The feed rate was 4 g/min or 12 g/min. Phosphorous pentoxide, particulates, GB, and other organophosphorous levels were also measured.

The combustion should take place according to the following reaction:

$$Na^+ \quad \underset{\underset{CH_3}{|}}{\overset{\overset{O}{\|}}{O\text{-}P\text{-}OCH(CH_3)_2}}{}^- \ + \ Na_2CO_3 \ + \ \frac{13}{2}O_2 \ \longrightarrow \ Na_3PO_4 \ + \ 5\ CO_2 \ + \ 5\ H_2O$$

The NaF, NaOH, and Na_2CO_3 salts are presumed to be retained in the melt, unaltered by the combustion process.

Increased feed rate, with constant temperature, melt depth, and melt composition, resulted in lowering P_2O_5 levels from 2.15 mg/333 l to 1.82 mg/ 333 l. It is theorized that burning elemental phosphorus in excess oxygen favors the exclusive formation of P_2O_5, while burning in limited oxygen reduces volatile oxides. Increased melt depth decreased P_2O_5 levels in both melt compositions. The mixed melt produced more P_2O_5 at higher temperatures while the 100% Na_2CO_3 melt decreased P_2O_5 at higher temperatures.

Previous incineration tests of phoshorous-containing compounds indicated that a mixed Na_2CO_3-Na_2SO_4 melt minimized P_2O_5 emissions. In tests, it was also observed that the pure carbonate melt not only increased P_2O_5 emissions, but also the total particulates. In some cases, the sodium carbonate melt expelled partially incinerated NaGB (23).

Distilled Mustard, HD--

Distilled Mustard was combusted in a melt consisting of 90% Na_2CO_3 and 10% Na_2SO_4. The quantity of salt mixture used produced an unexpanded bed depth of 30 cm. An air feed rate of 0.94 l/minute, increased the bed depth to about 60cm. Melt temperatures ranged from 915°C to 935°C.

The combustion should take place according to the following reaction:

$$ClCH_2CH_2-S-CH_2CH_2Cl \; + \; 2 \; Na_2CO_3 \; + \; 7 \; O_2 \; \longrightarrow$$

$$Na_2SO_4 \; + \; 2 \; NaCl \; + \; 6 \; CO_2 \; + \; 4 \; H_2O$$

No unincinerated mustard was detected in the off-gases, on the particulate filter, or in a sample of the melt. NO_x levels ranged from 4-15 ppm. SO_2 was less than 0.11 ppm, HCl ranged from 2.2-16 ppm, and CO levels were measured at 0.06%. Particulate emissions were 2.15 mg/333 l and consisted mainly of sodium carbonate, sodium sulfate, and sodium chloride. Destruction greater than 99.999997% was reported (23).

VX--

VX was combusted in the same type of melt used for distilled mustard. The temperature was maintained between 920-930°C. The combustion for VX should take place according to the following reaction:

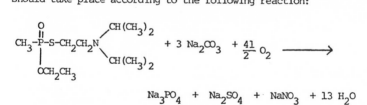

$$Na_3PO_4 \; + \; Na_2SO_4 \; + \; NaNO_3 \; + \; 13 \; H_2O$$

Unincinerated VX in the off-gases, on the particulate filter, or in a melt sample was negligible. NO_x levels ranged from 9-70 ppm, SO_2 was less than 0.14 ppm, CO levels ranged from 0.04-0.10%. P_2O_5 was measured at 856 ppm. Particulate data were incomplete. Destruction greater than 99.999988% was reported (23).

Lewisite, L--

Lewisite was combusted in the same type of melt used for distilled mustard. The temperature was maintained between 810-880°C. The combustion should take place according to the following reaction:

$$ClCH{=}CHAsCl_2 \;+\; 3\;Na_2CO_3 \;+\; 3\;O_2 \;\longrightarrow\; 3\;NaCl \;+\; Na_3AsSO_4 \;+\; 5\;CO_2 \;+\; H_2O$$

Less than 0.0053 mg/1000 l unincinerated Lewisite was detected. NO_x was reported at 15 ppm, CO levels ranged from 0.04-0.10%, and HCl was 11-15 ppm. Particulate arsenic emissions averaged 299 ppm. Residual arsenic in the melt ranged from 2.1-2.4 mg/gm (23).

GB—

The combustion of GB in molten salt should take place according to the following reaction:

$$\underset{\underset{CH_3}{|}}{\overset{\overset{O}{\|}}{F{-}P{-}O{-}CH(CH_3)_2}} \;+\; 2\;Na_2CO_3 \;+\; \frac{13}{2}O_2 \;\longrightarrow\; NaF \;+\; Na_3PO_4 \;+\; 6\;CO_2 \;+\; 5\;H_2O$$

Test runs using a melt of the same composition as that used for distilled mustard showed no detectable amounts of GB in the off-gas. Significant levels of P_2O_5 (as high as 955 ppm) and Na_3PO_4 (value not stated) were reported. Fluoride determinations were not made. Destruction of GB was calculated to be 99.9999985%.

Pesticides and Herbicides

A patent was assigned to Rockwell International Corporation in 1974 for a molten salt disposal process invented by Yosim et al. This process was designed for the ultimate disposal of organic pesticides and herbicides with negligible particulates and off-gas pollution. According to the process description, the organic pesticide and a source of oxygen are fed into a melt composed of sodium carbonate and from 1-25 wt% sodium sulfate. Temperature ranges are between 850-1000°C with an optimum range at 900-950°C (21).

Certain pesticides are completely combusted in the molten salt with an excess of oxygen. Other pesticides are partially oxidized in the melt and the remaining gaseous products are conducted into a second reaction zone where oxidation of any combustible matter still present is completed. Usually more than one zone is used to achieve complete combustion and ultimate disposal of the pesticide (21).

When chlorinated hydrocarbon pesticides are treated, NaCl is formed in the melt. Sodium phosphate is formed during the treatment of organic phosphate pesticides. Sodium sulfate is produced from the treatment of sulfur-containing pesticides. All of these inorganic compounds are retained in the melt. When the melt is no logner able to react with the pesticide, the salt is removed and fresh salt added (21).

Chlordane--

A few tenths of a gram of 50% chlordane ($C_{10}H_6Cl_8$) powder were added intermittently to a 85 wt% Na_2CO_3-15 wt% Na_2SO_4 melt maintained at 980°C. Air flow through an air inlet tube forced the pesticide through about 30.5 cm of molten salt in order to permit better contact between pesticide and melt. It was concluded that 99.9% of the pesticide was rapidly decomposed. Exit gases contained products of the reaction between carbonaceous material and the sulfate–primarily carbon dioxide and carbon monoxide. If the ratio of air to pesticide increased, the presence of carbon monoxide in the off-gas decreased. The levels of chlorinated hydrocarbons in the off-gas ranged from 10 ppm to 150 ppm. The higher range of hydrocarbon levels resulted when the air to pesticide ratio was relatively low. No chlorides were detected in the water scrubber when silver nitrate was added. Assay demonstrated that the melt retained about 80 wt% of the total chlorine.

Larger amounts of chlordane, contained in polyethylene bags at levels of 5-10 gm, were added to the melt through an alumina air tube which was open at both ends. The 10 gram quantities were substantially destroyed in about two minutes.

Liquid chlordane as a 72% emulsifiable concentrate was completely combusted in a continous feed system with excess air. Analysis of particulate filter and water scrubber fractions, used to trap any emitted organic chlorides indicated greater than 99.9% destruction of the pesticide by the molten salt.

Assay of the off-gas indicated levels of NO_x at less than 70 ppm, hydrocarbon emissions at less than 25 ppm, less than 0.1% CO, and 80% N (21).

Malathion--

Malathion is a representative organophosphorous-type pesticide which also contains the heteroatom sulfur. Ten polyethylene bags, each containing 5 gm of Malathion, were added to the melt at three minute intervals via an additional port. The closed port forced all emissions through the secondary combustor (described previously). Emissions were then passed through a glass wool particulate trap, through water scrubbers, and finally out of the system (21).

Benzene extracts of the particulates in the glass wool and water scrubbers were evaporated to dryness and analyzed for residues. Analyses reported 1.2 mg sulfur and 1.8 mg phosphorus. These figures represent 99.9% destruction of the pesticide. The melt analysis assayed 80 ± 15%

phosphorus. No CO or hydrocarbons were detected above the secondary combustion region. Monitoring of CO and hydrocarbon emissions indicated that 5 grams of sample were destroyed in about 30 seconds.

Grantham et al. combusted Malathion powder in the Rockwell International bench-scale combustion unit shown in Figure 2 (depicted earlier). A 900°C, 15-cm deep melt consisting of either Na_2CO_3 or K_2CO_3 was used. Destruction of the pesticide was reported greater than 99.99%. Although no residual pesticide was detected in the melt, traces of Malathion (within the threshold limit value) were sometimes detected in the off-gas. An increase of the bed depth to 25 cm markedly reduced pesticide levels, and decreased off-gas particulates (22).

Yosim et al. also used the bench-scale combustion unit in Figure 2 (depicted earlier) to combust Malathion dissolved in xylene. He reported better than 99.99% removal at 1000°C. No pesticide was detected in the melt, however, in one test a trace of Malathion was found in the off-gas.

Weed B Gon--

Weed B Gon consists of 17.8 wt% of the isooctyl ester of 2-4-D and 8.4 wt% of the isooctyl ester of silver. Six polyethylene bags, each containing 5 g of the herbicide, were added to a melt comprised of 90 wt% Na_2CO_3 and 10 wt% Na_2SO_4 maintained between 950°-1000°C. Analysis of benzene extracts indicated 0.77 mg residue of the 1.48 g chloride added to the system in the total emissions. Theoretically this represents a pesticide destruction of at least 99.96% (21).

Sevin--

Sevin also known as Carbaryl, 1 Napthyl N-methylcarbamate, contains 7 wt% nitrogen. Five gram packets of this typical carbamate pesticide were added to the same type of melt used for Weed B Gon. Analysis of benzene extracts indicated that 0.075 mg of 0.875 g of initial nitrogen remained. Destruction was reported to be 99.99%. Because of the rapid carbon-nitrate-nitrite reduction that occurs in the melt, nitrogen was not detected in the melt. Peak nitrogen emissions were 20 ppm (21).

DDT Powder and DDT-Malathion Solution--

Tests with the Rockwell bench-scale molten salt combuster were conducted for DDT powder and solutions of DDT and Malathion dissolved in xylene. This data is summarized in Table 2. These substances were combusted at 900°C in a 15-cm deep salt bed containing either Na_2CO_3 or K_2CO_3. The feed rate was 227-907 g/hr. Destruction of the pesticide was greater than 99.99%. Although no pesticides were detected in the melt, traces (0.3 mg/1000 l) were found in the off-gas (8).

2,4-D Herbicide-Tar Mixed Waste--

A mixed waste composed of 30-50% 2,4-D (an ester of dichlorophenoxy acid) and 50-70% bis-ester and dichlorophenol tars, was reported completely

TABLE 2

TYPICAL RESULTS OF DESTRUCTION TESTS WITH MALATHION AND DDT

Pesticide	Salt	Pesticide Destroyed %	Concentration of Pesticide in Melt (ppm)*	Quantity in Exhaust Gas (mg/m^3)*	TLV* of Pesticide (mg/m^3)*
DDT	Na_2CO_3	99.998	ND 0.05	0.3	1
DDT	K_2CO_3	99.998	ND 0.2	0.3	1
Malathion	Na_2CO_3	99.9998	ND 0.01	0.06	15
Malathion	K_2CO_3	99.999	ND 0.005	ND 0.4	15

*ND = not detected, one mg/m^3 = 4.4 x 10^{-4} grains/scf, TLV = threshold limit value.

destroyed in the Rockwell International pilot plant molten salt combustor at 830°C. No organic chlorides or HCl were detected in either the melt or off-gas. The waste was first diluted with ethanol to reduce viscosity (8).

Other Pesticides--

Since the molten salt process involves destruction of an organic compound, whether the initial step is partial or complete destruction, Yosim et al. postulated that a wide variety of organic pesticides may be destroyed with relatively minor treatment modifications. Possible candidates for molten salt destruction are DDT, dieldrin, heptachlor, aldrin, toluidine, the nitrile herbicides (trifluralin, 2,4,5-T dichlorobinil, MCPA), and the phosphorous containing insecticides (diazinon, disulfonton, phorate, and parathion) (21).

Real and Simulated Pesticide Containers--

Combustion tests were conducted on bench-scale and pilot plant levels for pesticide container materials comprised of paper, plastic, rubber, and a blend of these. Combustion was reported complete in all cases. The value of carbon monoxide in the off-gas was less than or equal to 0.2%. NO_x values were less than 65 ppm, and unburned hydrocarbons were detected at less than 30 ppm. In the tests with PVC, no HCl was detected in the off-gas.

The pilot plant combustor was used for the destruction of 700 kg simulated pesticide container material at a feed rate of about 30 kg/hr. Complete and rapid combustion was reported. The waste was composed of 53 wt% paper, 32 wt% polyethylene, 8 wt% PVC, and 7 wt% rubber. Less than 5 ppm wt% HCl, less than 2 ppm sulfur dioxide, less than 0.1% CO, and less than 0.1% hydrocarbons were detected in the off-gas. NO_x was about 30 ppm. Values of 5-12% O_2, 10-15% CO_2 and 76-78% N_2 were reported in the off-gas. The relatively small pesticide residues found in empty containers indicated it is possible a cooler melt (Na_2CO_3-K_2CO_3 eutectic, m.p. 710°C) could be considered.

Noncombustible materials such as glass and metal were also tested. Glass reacted completely in about 30 minutes at 900°C. At 1000°C, the glass reacted rapidly. Reaction of a metal (unspecified) was considerably slower, with surface corrosion to a depth of 1.75 mm in 8 hours at 900°C. It was recommended that molten salts not be used to completely disintegrate metal containers. A decontamination rinse by immersion for several minutes is an alternative (8).

Hazardous Organic Liquids

PCB's--

Current EPA regulations require incineration at 1200°C, and two seconds residence time, to ensure complete PCB destruction.

Molten salt PCB destruction was tested with the Rockwell International bench-scale unit. The PCB's studied (from voltage transformers), had an empircal composition of $C_{9.08}H_4Cl_4$. Tests were performed at temperature ranges of 700-980°C and air-PCB ratios from 90-230 wt% stoichiometric air. An initial melt of Na_2CO_3 with K_2CO_3 added to lower the melting point was used. Because NaCl was formed in situ by reaction of the carbonate with the chlorine of the PCB waste, the melt composition varied (7,22).

A gas chromatograph monitored PCB levels during molten salt combustion. Approximately 20% of the off-gas was scrubbed in benzene bubbles to absorb PCB. The samples were concentrated, mixed with freon, reevaporated, and mixed with exactly 1 cc of carbon disulfide. Duplicate and triplicate 60 μl samples of the carbon disulfide mixture were assayed in the gas chromatograph. The analytical methodology was verified by employing the same methodology with samples of the original PCB (Inerteen 70-30). As little as 100 ppb PCB was measured with the flame ionization detector gas chromatograph. PCB standards were diluted and used to establish chromatographic responses and evolution times for several PCB's.

Table 3 summarizes the result of the PCB tests. No PCB's (less than 70 μg/1000 l) were detected in any of the off-gas samples when all the Na_2CO_3 had been converted to NaCl. Destruction was estimated to exceed 99.99999%. However, it is necessary to maintain sufficient carbonate levels (about 2 wt%) and excess air in order to ensure complete combustion and maintain emission of NO_x, CO, CH_4, H_2, and unburned hydrocarbons at low, acceptable levels. The nominal residence time of the PCB in the salt was 0.25-0.50 seconds based on a flow-rate of 30-60 cm/sec through 15 cm of the melt (7,22).

Chloroform—

Rockwell International scientists have combusted chloroform in a molten salt reactor using both continuous feeding and bulk feeding techniques. Chloroform was fed to the bench-scale combustor at 227-907 g/hr. The temperature of the Na_2CO_3 melt was 850°C. Unreacted chloroform and HCl was not found in the off-gas. Greater than 99.999% destruction was reported. The same melt parameters were used for combustion in the pilot plant. Bulk quantities of chloroform were contained in metal canisters and plunged into the melt. This technique eliminated inconvenient, costly shredding equipment (7).

Three to four sealed Pyrex vials containing 160 ml chloroform were packed in cardboard tubes or sawdust and placed in 10-cm diameter metal canisters which were sealed. In some tests, the metal canister was punctured to allow melt access to the glass vials. The canisters were plunged into the melt. Vials in the punctured canister ruptured after less than 1 minute immersion in the melt with little noise. Vials in the unpunctured canisters ruptured noisily from 1-3 minutes after immersion into the melt (7).

Monitoring the off-gas with Drager tubes revealed less than 1 ppm hydrogen chloride, less than 1 ppm phosgene, and less than 5 ppm chloroform.

TABLE 3

PCB COMBUSTION TESTS IN SODIUM-POTASSIUM-CHLORIDE-CARBONATE MELTS

Temp °C	Stoichio-metric Air (%)	Concentration of KCl, NaCl in Melt (Wt%)	Extent of PCB* Destruction %	Concentration of PCB in Off-gas** (g/m^3)
870	145	60	ND 99.99995	ND 52
830	115	74	ND 99.99995	ND 65
700	160	97	ND 99.99995	ND 51
895	180	100	ND 99.99993	ND 59
775	125	100	ND 99.99996	ND 44
775	90**	100	ND 99.99996	ND 66

* ND = None detected; one g/m^3 = 4.4 x 10^{-7} grains/scf.
** Insufficient air present to completely oxidize PCB to CO_2 and H_2O, i.e., some CO, H_2, and CH_4 present in the off-gas.

Source: Rockwell International (1979).

Trace amounts of chloroform were found in all the benzene scrubber samples analyzed. However, the levels were near the detection limits. Chloroform destruction was estimated at 99.95% (7).

Perchloroethylene Distillation Bottoms--

Perchloroethylene (C_2Cl_4) is a common industrial solvent. During the refabrication of high temperature, gas-cooled nuclear reactor fuel, perchloroethylene is used to scrub condensible cracked and incompletely cracked hydrocarbons, carbon soot, and uranium-bearing particulates from off-gas. Eventually, the perchloroethylene degrades and requires distillation. Oak Ridge National Laboratory reports that the bottoms from this distillation contain a large number of complex polynuclear aromatic compounds. Although these compounds are variable and difficult to characterize, there is good analytical evidence that this waste contains carcinogens. Perchloroethylene distillation bottoms represent a type of hazardous waste for which complete destruction is necessary.

A slurry which contained 93 wt% perchloroethylene, 6 wt% organic degradation products, and about 1 wt% solids (carbon, silicon carbide, uranium, etc.) was feed by peristaltic pump into a Rockwell International bench-scale molten salt combustor. Since the heating value of the waste was less than 70 joules/g, kerosene was added to furnish the necessary heat to maintain the temperature (6,7).

Although experimental conditions such as temperature (850-950°C), stoichiometry (30-100 wt% air), and perchloroethylene-kerosene weight ratio (8-1) were varied, no peaks of organic material above background levels were found in any off-gas sample. Based on the analytical methods employed, it was concluded that any compounds in the off-gas from the combustion of perchloroethylene bottoms should not be in greater concentration than 0.5 mg/1000 l. As long as about 1% Na_2CO_3 remained in the melt, the concentration of HCl in the off-gas was less than 2 ppm. NO_x levels in the off-gas were 25 ppm. The particulate content of the exhaust gas increased with increasing time. This was expected since the particulates consisted mainly of NaCl formed from the vaporization of NaCl in the melt (6,7).

Trichloroethane--

Trichloroethane ($C_2H_3Cl_3$) is an industrial chemical which contains 80 wt% Cl. This chemical was combusted in the bench-scale unit at the rate of 227-907 g/hr.

This test was designed to define the amount of sodium carbonate which could be converted into NaCl without affecting the chloride scrubbing capacity of the melt. With as little as 2 wt% Na_2CO_3 remaining in the melt, only a negligible amount of trichloroethane was detected in the off-gas. It was estimated that 99.999% of the chemical was destroyed (6,7).

Nitroethane, Diphenylamine HCl, Monoethanolamine—

When the nitrogen-containing compounds monethanolamine (C_2H_7ON), nitroethane ($C_2H_5NO_2$), and diphenylamine HCl ($C_{12}H_{12}NCl$), were combusted in the Rockwell International bench-scale molten salt unit at 840-922°C, no unreacted material was detected and 99.99% destruction was estimated. Yet, these nitrogen-containing compounds did produce substantial amounts of NO_x. Further tests indicated that NO_x concentration could be sharply reduced by adjusting the air/waste fuel ratio. In theoretical 165 wt% (excess) air, monoethanolamine combustion produced 2200 ppm NO_x. This quantity was reduced to 220 ppm at 108 wt% theoretical air if combustion takes place in a semi-reducing environment. As nitroethane contains a nitro group rather than an amine, the NO_x emission was 16,000 ppm at 100 wt% stoichiometric air. Theoretically, increased reducing conditions could lower the NO_x concentration below 108 ppm (6,7).

Tributyl Phosphate—

Tributyl phosphate ($C_{12}H_{27}PO_4$) often used as a fire retardant, is difficult to burn in a conventional incinerator. Both pure and diluted tributyl phosphate (in 30% kerosene) were combusted at 900°C with 45 wt% and 28 wt% excess air. The high CO_2 and low CO and hydrocarbon concentrations in the off-gas indicated rapid consumption of the chemical. The CO_2 content of the off-gas measured 10-14%, CO, 0.5%, unburned hydrocarbons 20 ppm, and 30 ppm NO_x (7).

Hazardous Solids

The following hazardous solids were combusted in the Rockwell International bench-scale molten salt unit.

Rubber—

In tests conducted to study the production of low Btu gas from industrial wastes, rubber tire buffings were gasified at 920°C with 33 wt% theoretical air (percentage of air required to oxidize material completely to CO_2 and water). Since the buffings contained organic sulfur which would form Na_2S in the melt 6 wt% Na_2S was added to the sodium carbonate melt to simulate conditions and function as a catalyst to accelerate char gasification (18).

The CO content of the off-gas was considerably lower than from the oxygen-containing wastes that were also used in the tests (i.e., wood, nitropropane, and film). No significant amounts of H_2S or other sulfur-containing gases (less than 30 ppm) were detected in the off-gas.

Para-Arsanilic Acid—

Para-arsanilic acid was combusted at rates of 227-907 g/hr. No unreacted material was dected in the melt or off-gas in tests performed at 925°C. The main combustion product, sodium arsenate, was retained in the melt as expected. Thus the melt must also be considered as hazardous (7).

Contaminated Ion-Exchange Resins--

Ion-exchange resins are difficult to burn in a conventional incinerator and generally produce a smoky flame containing evolved hydrocarbons.
Tests performed on nuclear contaminated Dowex-1, and Powdex resins (styrene
divinyl/benzene cross-linked polymer with trimethyl amine anion or sulfonic
acid cation component) revealed rapid and complete destruction (7).

High-Sulfur Waste Refinery Sludge--

A high-sulfur (up to 15 wt%) waste refinery sludge was readily processed in the Rockwell International bench-scale molten salt gasifier. As
a useful by-product, a high quality, low-heating value (170 Btu/scf), low-
sulfur gas was produced (7).

HAZARDOUS WASTES DESTROYED BY THE FLUIDIZED BED PROCESS

Background

Many industrial uses have been proposed for fluidized bed systems
since this technique was suggested by C.E. Robinson about a century ago.
However, it was not until the late 1920s that the first commercial fluidized bed unit involving a gas-solids mixture and utilizing elevated
temperatures was installed by the petroleum refining industry (9). The
phenomenon of fluidization supports the transport of a large bulk of
catalyst from a reactor to a regenerator and back (10). In 1942, the
Standard Oil Company of New Jersey opened a fluidized bed petroleum
catalytic cracking plant in Baton Rouge, Louisiana (9). Since then,
fluidized solids technology has been firmly established as a useful and
valuable industrial operation (11).

After it was adopted by the petroleum industry, fluidized bed technology was successfully applied to many gas-solids operations in other
industries. Included in these applications were metallurgical processes
such as roasting sulfide ores and oxide ore reduction. Fluidized beds have
also been used extensively by the pharmaceutical and food industries for
rapid and intensive drying of powdery and granular materials (11).

Because fluidized bed incineration incorporates both waste disposal
and energy recovery features, there has been much recent interest in the
technology as it relates to coal gasification, electric power generation,
and boiler units. The state-of-the-art of these energy efficient applications in 1977 was summarized at the International Conference on Fluidized
Bed Combustion (32).

One of the first applications of fluidized bed technology for the
incineration of carbonaceous waste material was in the pulp and paper
industry. Research on this application was begun at the Columbus Laboratories of Battelle Memorial Institute in the late 1950s. This culminated
in the erection of a Container-Copeland commercial installation at the
Carthage, Indiana mill of the Container Corporation of America in 1962.
Fluidized bed combustion is especially useful in the combustion of spent

pulping liquor. The spent liquor contains sodium-sulfur compounds of varying degrees of oxidation in association with organic matter extracted from wood. In 1977, there were more than 25 commercial fluidized bed installations in the pulp and paper industry (9).

Dorr-Oliver entered the fluidized bed incineration field in 1960 with a small unit for handling approximately 10 kg/h of primary sludge. This was followed by a commercial sized unit built at Lynwood, Washington, in 1962. Since then, commercial fluidized bed units have incinerated many different kinds of industrial waste. These include oil refinery waste, primary and activated sludges, and carbon black waste (33).

Although fluidized bed technology is established in many industries, it has not been extensively used for the ultimate disposal of hazardous waste. In 1978, Systems Technology Corporation (Systech) in Franklin, Ohio, evaluated fluidized bed combustion in the destruction of methyl methacrylate and phenol waste for the U.S. Environmental Protection Agency. Incineration of each waste was accomplished with high efficiencies in the combustor (34). Other investigators have evaluated fluidized bed combustion of a limited number of hazardous wastes. The following material will detail new and emerging treatment systems for the combustion of hazardous waste by fluidized bed technology. Both bench scale and pilot plant systems will be discussed.

Chlorinated Hydrocarbons

There have been several different types of tests in which chlorinated hydrocarbons, e.g. polyvinyl chloride (PVC), have been combusted in fluidized bed incinerators. Polyvinyl chloride (PVC) is one of the more widely used plastics. The monomer of PVC, vinyl chloride, has two carbon atoms, three hydrogen atoms, and one chlorine atom. The chlorine content is 45 wt% of pure PVC.

PVC is rarely pure and usually contains fillers, stabilizers, plasticizers, flame retardants, and other chemicals. The major products of PVC combustion are carbon monoxide, carbon dioxide, and hydrogen chloride. Free chlorine is not a product of PVC combustion, but trace quantities of carbonyl chloride and vinyl chloride have been reported. Fifty-six other volatile products have been observed in small amounts (35).

Chlorinated Hydrocarbons/Bench Scale Processes

Plastic Waste Combusted with Coal--

A study was done on combustion of PVC mixed with coal in a fluidized bed. The effectiveness of dolomitic limestone, aluminum oxide, and silica sand in removing hydrochloric acid was also investigated. Ragland and Paul used a 9-cm diameter, 0.3-m tall, quartz-lined fluidized bed. The tubular combustor wasd flanged to a 0.3 m long freeboard (area above the fluidized bed) tube and a 12-cm diameter cyclone particulate collector. A water cooled pneumatic feed system was used near the bed. The distributor plate was porous, sintered bronze. For start-up, a propane-air mixture was used to preheat the bed. After preheating, the propane was shut off, and the inlet air was preheated to approximately 90°C by means of a coiled tube in the freeboard. Flashback in the propane-air mixture was controlled with a flash arrestor, and the plenum chamber (below the distributor plate) was filled with steel wool (35).

Preliminary experiments showed that thin film PVC sheet, when shredded to 0.3 cm, could be readily fluidized in a sand bed without segregation. Blends of 5 to 30 wt% shredded PVC with Montana subbituminous coal containing 25% moisture, 35% carbon, 10% ash, and 0.6% sulfur were used in the feed mixture. The particle size of the coal was 8-20 mesh. The PVC, which contained some filler, was ground and sieved to the same size range. The mixture had a 17.5 wt% chlorine content.

The efficiency of dolomitic limestone (calcium carbonate with calcium magnesium carbonate), aluminum oxide, and silica sand (size range 8-20 mesh) as chlorine scrubbers was investigated. These materials have melting points well above the bed temperature. The major source of chlorine was the PVC. The chlorine content of the coal, aluminum oxide, and silica sand was negligible. Neutron activation revealed 0.4% chlorine in the dolomite. Bed depth was approximately 7.6 cm.

Combustion tests were conducted using 5, 20, and 30 wt% PVC-coal blends with the three different bed materials. The temperature averaged 840°C. Good fluidization was obtained with a flow of 0.25 m³/minute. The PVC/coal mixture was injected with a pulse of air about every 5 seconds. Three hundred grams of feed were burned every 7-10 minutes.

The combustion gases from the fluidized bed were sampled at the cyclone exit with a modified EPA method 6 sampling train. A 1-cm diameter quartz probe electrically heated to 200°C was used. A pyrex wool plug in the end of the probe removed particulates before the sample was drawn into four midget impingers in an ice bath. Three impingers contained dilute NaOH solution for absorbing HCl. The last impinger removed moisture. A Teldyne Model 980 flue gas analyzer was used to pump the gas sample and to determine the oxygen and gaseous combustibles in the flue gas. The impinger solution and probe-plug wash were titrated with mercuric nitrate to determine the HCl concentration.

Chlorine emissions for the three different beds and at different PVC concentrations are shown in Figure 8. The concentration of HCl in the flue gas was 500 ppm with 5 wt% PVC, using silica sand and aluminum oxide beds. A dolomite bed decreased this concentration to 135 ppm. The total possible HCl emission was calculated at 900 ppm for 5 wt% PVC. The ability of dolomitic lime stone to scrub HCl is due to the reaction between calcium oxide and HCl to form calcium chloride. The formation of magnesium chloride could also be effective in removing HCl. Aluminum oxide was not nearly as effective as dolomite, and aluminum chloride was apparently not formed. The spent silica sand and aluminum oxide beds contained little chlorine (Figure 9). The retention of chlorine in these systems occurred primarily in the cyclone ash.

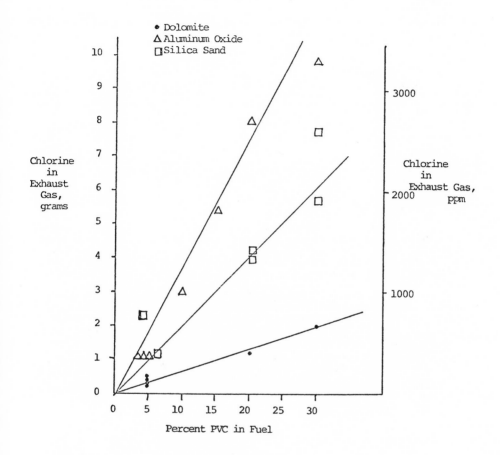

Figure 8.   Chlorine Emissions in Flue Gas

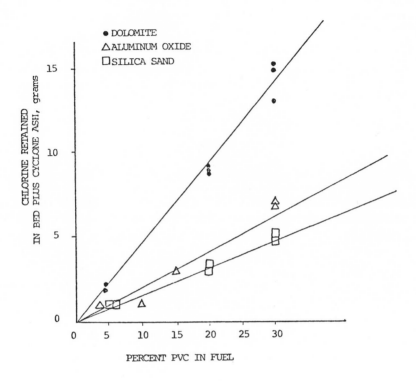

Figure 9.   Chlorine Adsorbed in Bed Material and Cyclone Ash

Although the prime chlorine scrubbing action of dolomitic limestone is probably due to its calcium carbonate content, the relative efficacy of limestone versus dolomitic limestone was not compared. The role of metal oxides and calcium oxides in the coal ash in chlorine scrubbing is apparently significant but not as yet understood. The effectiveness of HCl removal if high sulfur coal is used should also be investigated. The paper contained no other emission data (35).

PVC Waste in a Tilting Fluidized Bed—

Kamino et al. (36) used a pyrolytic bench-scale process in an attempt to convert a plastic mixture containing PVC into fuel gas. To prevent formation of harmful gases and lower corrosion of the processing equipment, the waste was first dechlorinated in a standard fluidized bed with sand as the fluidizing medium and nitrogen as the fluidizing gas. The bed was maintained at 220-380°C (the temperature range in which PVC gives off HCl). As time elapsed, the plastic material cohesed with the sand, and the bed lost fluidity.

To eliminate this problem, the standard fluidized bed was replaced with an inclined fluidized bed (Figure 10). In the inclined bed, accumulation of material would be prevented by the outward movement of sand and plastic material. Tilting the bed does not require additional mechanical operating parts.

The inclined fluidized bed was made up of a liquidized gas flow area, a dispersal plate, and a reaction area. The fluidizing medium was sand with an average granular diameter of 1.2 mm. Propane gas functioned dually as the fluidizing gas and primary combustion source.

Sand, from the sand storage area, passed through a hot blast stove, became heated, and joined the plastic waste before they both entered the gas flow area of the combustor. Dechlorination was essentially completed in the reaction area. Residual chlorine gas from the process passed to an alkali tank for neutralization and venting to the atmosphere (36). Many engineering details were not supplied in this article.

Chlorinated Hydrocarbons/Pilot Plant Studies

Copper Wire Insulated with PVC—

A patent for a continuous fluidized bed process for removing insulation from copper wire or other materials was assigned to the Cerro Corporation of New York, New York. The process is particularly suitable when compared to stripping or burning. Stripping is unsatisfactory for fine wire, while burning oxidizes copper (37). A schematic of the combustor is shown in Figure 11.

The polyvinyl chloride insulated wire was pre-chopped in a shearing machine from 681 kg bales into 15 cm pieces and fed via conveyer through a water seal into a decomposition chamber. The decomposition chamber was circular in cross-section and divided by a hearth into an upper and lower

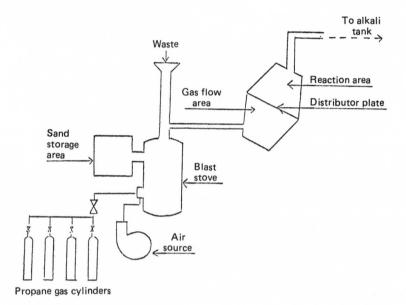

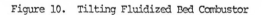

Figure 10.   Tilting Fluidized Bed Combustor

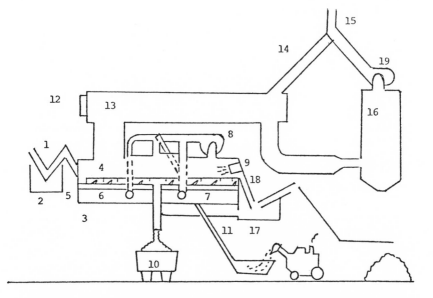

.Figure 11.   Schematic of a Continuous Fluidized Bed Process for
Removing Insulation from Copper Wire. (See legend below for details)

1. Feed wire conveyer
2. Water seal
3. Decomposition chamber
4. Advancing mechanism
5. Hearth
6. Pyrolysis section
7. Decomposition section
8. Fluidizing air fan ending in
   burners in pyrolysis and
   decomposition sections
9. Pilot burner
10. Collection area for calcium
    chloride and calcium carbonate
    discharged from bed

11. Feeder for bed material
12. Afterburners
13. Afterburner chamber
14. Air intake from afterburner
15. Gas discharge from scrubber
16. HCl scrubber—dust from the
    scrubber passes to the quenching
    tank and then to 10 (not shown)
17. Quenching tank
18. Clean wire comes out, calcium
    chloride and calcium carbonate
    transported to 10 (not shown)
19. Induced draft fan

section. The lower section of the decomposition chamber was divided into a pyrolysis section and a decomposition section (Figure 11). The hearth had many openings so that the upper section of the decomposition chamber was connected with the two lower sections.

A burner in the pyrolysis section of the lower chamber maintained a 310° C operating temperature with a reducing atmosphere. A burner in the decomposition section of the lower chamber maintained temperatures in the 640°C range with adequate oxygen for complete combustion.

The fluidizing medium, calcium carbonate (10-20 mesh) was transferred from a storage area to the decomposition section of the lower chamber and was heated to about 310-640°C. Fluidizing air was supplied by a fan to the two burners in the lower chamber. Heated calcium carbonate was fluidized by discharge from the burners.

Cut feed entered the upper chamber over the pyrolysis section and was pulled over the hearth by an advancing mechanism. The waste was subjected to pyrolysis and partially decomposed.

The advancing mechanism consisted of a rake assembly and a ramp assembly. Both the rake assembly and ramp assembly are movable independently or in unison with the advancing mechanism. A ramp assembly comprises a base member with cam blocks positioned in a track mounted at the sides of the furnace. The rake assembly consisted of parallel side bars connected by spaced cross rake members that had downward projecting tines (37).

As feed was moved along the hearth by ramp and rake, it was surrounded by heated particles of calcium carbonate and decomposition was completed. Some of the calcium reacted with the chlorine or hydrochloric acid produced by waste decomposition. A pilot burner located in the upper chamber provided mixing and safety ignition to prevent a build-up of a combustible atmosphere.

The clean copper wire left the decomposition chamber through an exit water seal which acted as a quench for the wire scrap. The wire was collected in a receptacle for further handling. The calcium salts were trapped in a weir and discharged into a receptacle (Figure 11). The product collected from the receptacle usually contained about 12% calcium chloride and 88% calcium carbonate. Calcium chloride is a useful by-product of the process, that can be used for snow removal and road building.

Because gases and smoke generated during decomposition in the chamber may contain small amounts of chlorine and carbonaceous material, the gases passed into an afterburner where all of the remaining combustible products were consumed. The afterburner, which operated between 760-1090°C, included a burner and an excess air inlet (Figure 11).

Since the burned gases from the afterburner may still contain chlorine as well as the products of combustion, these gases were passed through an HCl scrubber. The scrubber was a conventional bubble-plate type, with the

water on the plates containing calcium carbonate so that any chlorine in
the gas was reacted to form calcium chloride. The scrubber also removed
any dust such as calcium carbonate. The gases then passed from the top of
the scrubber into an induced draft fan that provided the pressure differen-
tial to cause gas flow through the system (Figure 11). The system was
designed to process 1362 kg/h of material (37).

Chlorinated Hydrocarbon Waste with High Chlorine Content
   Pilot Plant Trial--

   Chlorinated hydrocarbon impurities can occur as waste in the manufac-
ture of plastics, herbicides, solvents, and paints. A chlorinated hydrocarbon that contains as much as 70% chlorine (calorie value, 2,500-3,000
kcal/kg) can support combustion in a high performance incinerator. When
the chlorine content is increased above 70%, additional fuel must be
supplied to bring the total calorific value of the waste up to about 2,500
kcal/kg to maintain combustion. A pilot plant trial was done to see if a
chlorinated hydrocarbon waste containing 80% chlorine could support combustion without additional expensive fuel in a fluidized bed combustor (38).

   The reactor had an 460-mm internal diameter with a bottom section
tapering to a 300-mm diameter at the distributor, and was constructed of
refractory material cast into a mild-steel shell. The Type 310 distributor
had four 64-mm diameter bubble caps, each having six 9.5-mm diameter
radially drilled, equally spaced holes.

   The bed consisted of 100-150 kg of 6-mm building sand that was fed to
the reactor from a pressure-sealed variable-speed table feeder. Compressed
air (1 m/sec) was the fluidizing medium. After the windbox was fired and
the bed raised to the desired operating temperature, a metered flow of
waste was injected (with atomizing air) directly into the center of the
fluidized bed at a point 35 mm above the distributor. The Type 310 stainless steel injection nozzle was 6 mm in diameter and flatted at the end to
form a fan-tail spray. When auxiliary fuel was needed (in one test run)
distilled oil was injected in the same nozzle. The flow rate of the waste
was adjusted arbitrarily to give an off-gas oxygen content of about 4%.

   Off-gases from the reactor were passed through a 230 mm-diameter high-
efficiency cyclone and then through two packed absorption columns where
they were scrubbed countercurrently with a soda ash solution to remove both
hydrogen chloride and chlorine. The absorption columns were packed with
ceramic Raschig rings. After scrubbing, the gases were monitored for
oxygen content and analysed using Drager tubes for chloride content. They
were then vented to the atmosphere. Samples from the scrubbers were also
analysed for chloride content. Completeness of combustion and hydrogen
chloride formed were determined by calculating a chlorine mass balance over
the system.

   Only 85% of the chlorine content of the waste liquor was recovered in
the 1000°C run. This indicates that complete combustion was not achieved.
Fuel oil was added to raise bed temperatures to 1,100°C. Almost all (99%)
of the estimated chlorine, evolved as HCl and Cl_2 was recovered in the

1,100°C run. As the fuel oil was not added directly with the waste, the run was considered autogenous.

An alternative to scrubbing the exhaust gases before they are vented to the atmosphere is water-absorption of hydrogen chloride to produce acid for recycle. Maximum formation of hydrogen chloride requires combustion at temperatures around 1,500°C and steam injection. In this case, the use of a fluidized bed would not offer many advantages over a standard high-performance burner (38). Specific chlorinated compounds were not cited in this paper.

Chlorinated Hydrocarbons/Bench to Pilot Scale-up

PVC Waste Generated at Rocky Flats Plant of Dow Chemical/Bench Scale--

The Rocky Flats Plant (RFP) of Dow Chemical used stationary grate incineration (800-1000°C) to convert bulky combustible residues into a suitable form for recovering plutonium by aqueous chemistry. Corrosion of the incinerator and off-gas scrubbing system and damage to the incinerator refractory lining caused frequent shutdowns for maintenance and repair. The incineration system also produced plutonium oxide which was difficult and expensive to recover. Since PVC accounts for about 40% of the RFP waste mixture, hydrogen chloride (HCl) generated in the combustion of poly-vinyl chloride (PVC) plastics was considered to be the major source of corrosive gas. To eliminate these problems, fluidized bed combustion with a sodium carbonate bed was used for in situ neutralization of HCl generated by the PVC as it decomposed (39).

Feasibility studies were carried out in a 6.3-cm diameter, 81.3-cm long quartz tube. Materials subject to HCl corrosion were eliminated. All vessels and gas transfer lines were constructed of HCl corrosion resis-tant quartz, Pyrex, or polyethylene. Several equipment arrangements were used to promote greater efficiency. An early arrangement utilized a static bed of Al_2O_3 balls for gas preheating with a 15-cm diameter solids disen-gagement section at the top of the reaction tube to separate solids from the off-gas stream. The fluidized bed was composed of alumina granules (Figure 12).

In a later design, the solids disengagement section was replaced by a Pyrex cyclone to improve the separation of solids from the off-gas stream (Figure 13). A fluidization gas preheater, two stage caustic scrubbers, and a fixed sodium carbonate bed for HCl neutralization, plus a catalytic afterburner were also added. Catalytic afterburning at 500°C offered the potential for an incineration system which did not need refractory lined equipment. The types of catalysts used will be discussed later. All gas transfer lines were constructed of Pyrex and were connected to the vessels by standard tapered glass fitting. The gas preheater was a 3,000 watt Chromalox commercial heater with piping to permit the use of air, argon, and oxygen gas mixtures for bed fluidization. The caustic scrubbing system consisted of two, 2,000-ml glass cylinders partially filled with 1.3-cm polyethylene Raschig rings and 1% sodium hydroxide solution (NaOH). Alumina balls were added as a roughing filter to eliminate packing of fine

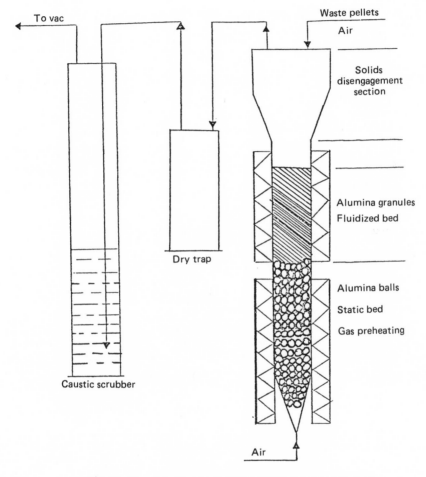

Figure 12.   Dow Chemical Rocky Flats Laboratory-Scale
Combustor.

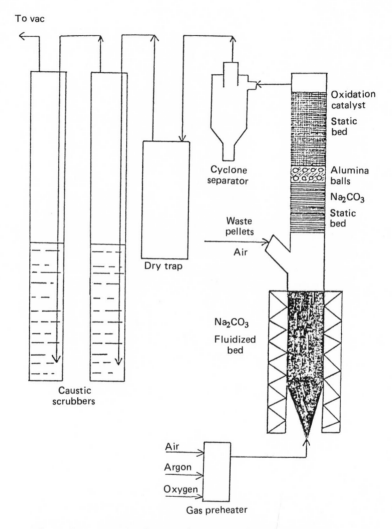

To vac

Oxidation
catalyst
Static
bed

Alumina
balls
Na_2CO_3
Static
bed

Cyclone
separator

Waste
pellets

Air

Dry trap

Na_2CO_3
Fluidized
bed

Caustic
scrubbers

Air

Argon

Oxygen

Gas preheater

Figure 13.   Dow Chemical Rocky Flats Laboratory Scale
Combustor: Intermediate Design.

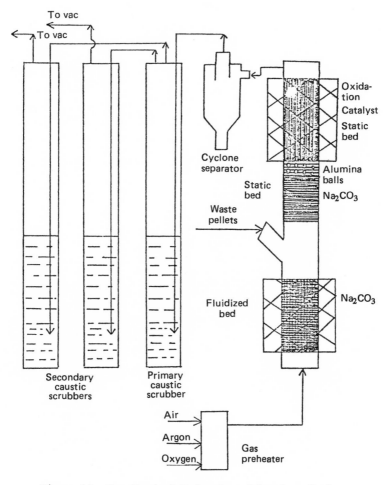

Figure 14.   Dow Chemical Rocky Flats Laboratory-Scale
Combustor: Advanced Design.

particles in the catalytic burner. Most incineration tests were made in
the unit shown in Figure 13. The fluidized bed was sodium carbonate.

The final design requirements are shown in Figure 14. A new quartz
incinerator vessel with 1000-mesh gas distribution screen improved bed
fluidization. The oxidation catalyst bed was enclosed in a furnace for
improved temperature control and off-gas combustion efficiency. Another
caustic scrubber vessel was added and off-gas piping was modified to
create a primary and secondary scrubbing system to prevent loss of HCl from
the system. The design was utilized for the last five incineration studies.
This fluidized bed was composed of sodium carbonate (39).

Waste was added as pellets to assure intimate mixing with Na_2CO_3 and
ease of feeding. The waste materials, PVC, polyethylene, paper, and
surgeons' gloves, were manually cut into pieces no larger than 1.27 cm.
They were then weighed and mixed into the correct proportions for a stan-
dardized pellet. The mixture was then pelletized by a batch extrusion that
held the materials together in a pellet approximately 6.3 cm long and
1.2 cm in diameter.

Twenty grams of pellets were placed in the bottom of a 200-ml crucible,
covered with powdered Na_2CO_3, and heated for three hours in a 600°C muffle
furnace. The residue in the crucible was dissolved in hot water and the
solution analyzed for chloride content. The average chloride content was
21.85 wt%.

Decomposition of waste pellets in the combustor was the major consideration in early bench-scale tests, and neither sodium carbonate for HCl
neutralization nor catalytic afterburning were used. Air was metered
through a bed of heated 0.7 cm-diameter alumina balls for preheating and gas
distribution. The heated air fluidized an alumina granule bed. When the
bed temperature reached 600-650°C, waste pellets were dropped in the top of
the disengagement section for decomposition in the fluidized bed. Decomposition tars and soot were retained in the dry trap and caustic scrubber.
A propane-oxygen flame was introduced above the fluidized bed in an attempt
to burn the waste decomposition gases. Data from the early combustion
tests were not given, but the fluidized bed combustion technique was considered feasible for waste from the Rocky Flats plant.

Emphasis was shifted to the goal of complete HCl neutralization and
clean-up of the off-gas stream with the modified units shown in Figures 13
and 14. Air or argon-oxygen was metered through a Chromalox preheater to
help maintain an operation temperature of 600°C within a fluidized bed of
Na_2CO_3. Waste pellets and combustion air were introduced through a side
port directly above the fluidized bed. The air and decomposition gases
were drawn through the system by a vacuum applied to the scrubbers. Decomposition gases were passed through a static bed of sodium carbonate in
order to react any HCl gas that might escape from the fluidized Na_2CO_3
bed. The flow continued into an oxidation catalyst bed to promote combustion of the waste decomposition gas. Air drawn in the side inlet
provided additional oxygen for combustion within the catalyst bed. When
the cyclone separator proved relatively successful in removing the small

amounts of solids that escaped in the incinerator, the use of dry trap was discontinued (39).

Off-gas was drawn through the cyclone into the scrubbing system. In the final scrubber design, all of the off-gas passed through a dip tube to the bottom of the scrubbing vessel. After a thorough mixing with caustic solution in the Raschig ring-filled column, the stream was split as it left the vessel and entered the house vacuum system (Figure 14).

Off-gases were sampled by attaching an evacuated 10-cm gas cell to the off-gas line and drawing the sample into the cell. The gases contained in the cell were then analyzed by infrared spectrophotometry. No chlorinated hydrocarbons were detected in the off-gases. Sodium carbonate probably neutralized HCl by the following reaction: (39)

$$Na_2CO_3 + 2\ HCl \longrightarrow 2\ NaCl + H_2O + CO_2$$

In attempts to increase the degree of neutralization of sodium carbonate on HCl, three different methods of neutralization were evaluated:
- pelletization of sodium carbonate with combustible PVC wastes to insure intimate contact during decomposition
- decomposition of waste PVC pellets in a fluidized bed of sodium carbonate granules (the method used in the initial studies)
- decomposition of PVC waste pellets in a fluidized bed composed of unreactive material (such as sand), and passing the gases produced in the reaction through a static sodium carbonate bed

Each method offered advantages, but each proved to be only partially successful when used alone. The degree of neutralization that was obtained by each method was measured by the amount of chloride retained in the fluidized bed after combustion. As seen from the data in Tables 4, 5, and 6, each method resulted in bed retentions of approximately 45-55% chloride.

When it became obvious that no single method could achieve 100% HCl neutralization, combinations of all three methods were evaluated. This resulted in improved chloride retention in the bed (Tables 7 and 8). The best results, 97.2% chloride retention, were seen when all three methods were combined (Table 9) (39).

Combustion efficiency of a Na_2CO_3 fluidized bed was expected to decrease as the bed NaCl concentration increased. One run was stopped to analyze the Na_2CO_3 beds, then continued to evaluate extended run efficiency. As expected, the percent chloride retained in the lower bed decreased approximately 6% during the second half of the run. When the run was stopped, the concentration of NaCl in the lower bed was 22%. The upper bed retained more chloride in the second half of the run. An upper, static bed is necessary for the best possible chloride retention in extended runs (39).

Waste pellet feed rates were examined as another variable that might affect HCl neutralization. The incinerators had no refinements for cooling

TABLE 4
HCl NEUTRALIZATION POTENTIAL
FLUIDIZED LOWER BED-ALUMINUM OXIDE
STATIC UPPER BED-NONE
WASTE ADDED AS PVC-SODIUM CARBONATE PELLETS

CHLORIDE RETENTION (wt%)					
Run No.	Total Recovery	Lower Bed	Upper Bed	Catalyst Bed	Scrubber	Lost
25	88.0	46.6	None Used	24.8	16.6	12.0

TABLE 5
HCl NEUTRALIZATION POTENTIAL
FLUIDIZED LOWER BED-SODIUM CARBONATE
STATIC UPPER BED—NONE
WASTE ADDED AS PVC PELLETS

CHLORIDE RETENTION (wt%)					
Run No.	Total Recovery	Lower Bed	Upper Bed	Catalyst Bed	Scrubber	Lost
14	60.6	49.4	None Used	No data	11.2	39.4

TABLE 6
HCl NEUTRALIZATION POTENTIAL
FLUIDIZED LOWER BED-ALUMINUM OXIDE
STATIC UPPER BED-SODIUM CARBONATE
WASTE ADDED AS PVC PELLETS

CHLORIDE RETENTION (wt%)					
Run No.	Total Recovery	Lower Bed	Upper Bed	Catalyst Bed	Scrubber	Lost
18	73.8	0	55.4	No Data	18.4	26.2

TABLE 7
HCl NEUTRALIZATION POTENTIAL
FLUIDIZED LOWER BED-SODIUM CARBONATE
STATIC UPPER BED-SODIUM CARBONATE
WASTE ADDED AS PVC PELLETS

		CHLORIDE RETENTION (wt%)				
Run No.	Total Recovery	Lower Bed	Upper Bed	Catalyst Bed	Scrubber	Lost
20	92.0	84.3	7.1	No data	0.6	8.0
21	90.8	74.3	15.3	No data	1.2	9.2
22	90.5	76.6	12.6	No data	1.3	9.5
Average	91.1	78.4	11.6	—	1.0	8.9

TABLE 8
HCl NEUTRALIZATION POTENTIAL
FLUIDIZED LOWER BED-SODIUM CARBONATE
STATIC UPPER BED-SODIUM CARBONATE
WASTE ADDED AS PVC PELLETS
(% SODIUM CARBONATE UTILIZATION)

		CHLORIDE RETENTION (wt%)					
Run No.	Sodium Carbonate Utilization (%)	Total Recovery	Lower Bed	Upper Bed	Catalyst Bed	Scrubber	Lost
24a	0-13	96.9	72.8	20.2	2.7	1.2	3.1
24b	13.0-22.2	95.5	66.6	23.1	1.5	4.3	4.5

TABLE 9
HCl NEUTRALIZATION POTENTIAL
FLUIDIZED LOWER BED-SODIUM CARBONATE
STATIC UPPER BED-SODIUM CARBONATE
WASTE ADDED AS PVC-SODIUM CARBONATE PELLETS

		CHLORIDE RETENTION (wt%)				
Run No.	Total Recovery	Lower Bed	Upper Bed	Catalyst Bed	Scrubber	Lost
19	97.2	90.9	5.8	No Data	0.5	2.8
26	95.3	88.7	5.9	0.4	0.3	4.7
27	96.6	91.6	1.8	1.0	2.2	3.4
28	94.8	87.2	0.2	7.0	0.4	5.2
29	91.8	87.8	1.4	0.3	2.3	8.2
Average	95.1	89.2	3.0	2.2	1.1	4.9

the upper bed, and the temperature increased in the upper bed when hydrocarbons in the flue gas combusted. Feeding rates were restricted when the
upper bed temperature exceeded 800°C. Normal feed rates below 800°C were
from 1.36 to 2.71 g/min. There was no obvious relationship between feed
rates and HCl neutralization when the temperature of the catalyst bed was
below 800°C.

During the course of the investigation, the quantity of HCl predicted
by calculation was consistently not recovered by rigorous analytical procedures. The possibility that HCl was lost during the pelletization process
was ruled out when a test run with unpelletized PVC also resulted in incomplete recovery. The cause of this problem was not detected. The possibility that chloride was lost through chlorinated hydrocarbon aerosol formation
was considered but not investigated.

These studies all used partial pyrolysis conditions rather than true
combustion. Separate runs were made with air, argon, and argon-oxygen as
the fluidizing medium. Residues in the incinerator were analysed for ash
content after each run. The average bed ash retentions were 9.5% for
argon, 5.3% for the argon-oxygen mixture, and 3.9% for air. High ash
retention in the argon-oxygen fluidized bed was accomplished before the
gases reached the catalyst section of the unit. High ash retention
indicated more efficient destruction in the bed with reduced need for above
bed combustion and scrubbing. Limited oxygen also reduced refractory oxide
formation in the bed and limited above bed burning. Optimum oxygen concentration appears to be between 10-15%, but the authors indicated a need for
more studies on gas ratios. There was no difference in bed fluidization
with any gas or gas mixture.

The need for open-flame afterburning was eliminated by adding an oxidation catalyst to the unit. Open-flame after-burners of incinerator flue
gas require 1,000-2,000°C for complete combustion of trace hydrocarbons.
The minimum temperature required for the catalytic after-burner is about
400 to 600°C. The lower temperature allows a wider variety of materials
which would be satisfactory for construction of the after-burner. Refractory lined equipment and costly re-bricking are eliminated.   The oxidation
catalysts will completely combust hydrocarbons at reduced oxygen concentrations. As there is no need for a large volume of oxygen in the afterburner, there is a decrease in the total volume of flue gas cleaned in the
HEPA (high efficiency particulate air) filter (39).

Six oxidation catalysts were tested for use in the incinerator afterburner. They were:

1. Shell 105X—Shell Chemical Company, 0.25 cm-diameter pellets,
    Fe_2O_3 with 21% potassium as a promoter
2. Zeolon 200H—Union Carbide, Norton Chemical Process Division
    0.12 cm-diameter pellets, 10% silicone oxide on a sodium (1%)-
    alumina support
3. ZR-0304T—Harshaw Chemical Company, 0.25 cm-diameter pellets,
    zirconium oxide (98%) mixed with alumina (2%)

4. CR-0211-Harshaw Chemical Company, 0.25 cm-diameter pellets, chromium oxide on alumina
5. Linde 13X molecular sieve—Union Carbide Company, Norton Chemical Company, Norton Chemical Process Division, 0.12 cm-diameter pellets (as shipped, H_2O 1.5 wt%)
6. Grace CAT 908-Davison Chemical Company, 0.7 cm-diameter balls, copper and magnesium oxides with alumina base

An oxidation catalyst used in place of incinerator after-burner in this fluidized bed system should exhibit good resistance to HCl attack and oxidation efficiency at low temperatures. To aid in selecting a suitable catalyst, a series of tests were made. The catalysts were contacted with a humid HCl gas stream at 500°C for two hours. Oxidation efficiencies were evaluated with a methane-air mixture.

Three of the catalysts, Harshaw CR-0211T, Linde 13X, and Grace CAT 908, exhibited almost 100% efficiencies from the beginning. The Linde 13X molecular sieve gained weight during HCl treatment, indicating HCl adsorption. It lost weight during oxidation tests, indicating HCl desorption. The Grace CAT 908 ha a large weight loss during oxidation tests without a similar weight gain during HCl treatment. This indicates that part of the catalyst was driven off as a chloride compound. The other catalysts exhibited a gradual increase in efficiency as a function of time. They were attacked by HCl but regenerated to some extent because of the oxidation conditions. The Grace CAT 908 and Harshaw CR-0211 exhibited almost 100% efficiency at 400°C. Linde 13X and Harshaw ZR-0304T required 500°C, and Norton Zeolon 200H and Shell 105X required 699°C for 100% efficiency. Of all the catalysts tested, Harshaw CR-0211 T (chromium on alumina) exhibited the best combination of resistance to HCl attack, oxidation efficiency, and temperature requirements.

Pilot-Plant Combustor at Dow Chemical—

Data from the bench-scale tests were used to design a fluidized bed pilot plant incinerator for evaluating process variables and obtaining process design information. A low-speed, cutter type shredder, unit capacity 10 m/hr, shredded material for incineration. Waste was shredded and mixed to provide a known composition of 44% PVC, 28% paper, and 28% polyethylene (40).

The pilot-plant incinerator was designed to combust waste at 4.5 kg/hr. A flow diagram is shown in Figure 15. The unit can operate with a single fluidized bed or with two fluidized beds in series. The lower (primary) bed diameter is 35.6 cm. The upper (secondary) bed diameter is 40.7 cm. The vessel diameter above the second bed increases to 61 cm and results in a gas velocity reduction and return of some entrained particles to the bed. Gas exits through a 10.2 cm diameter cyclone that removes more entrained solids. An overflow tube and air-jet ejector convey upper bed material to the lower bed.

Shredded waste was fed from a hopper by a chain conveyor that regulated waste feed rate. Waste then passed through a constant-pitch,

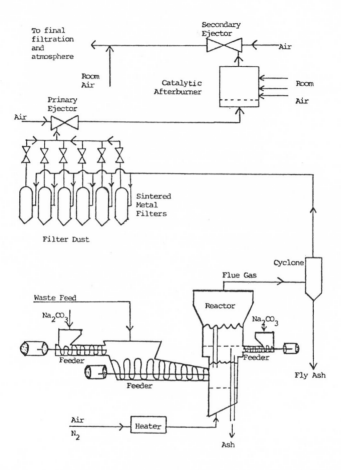

Figure 15.   Flow Diagram for Pilot Plant Fluidized Bed Incinerator

tapered-screw conveyor that introduced feed to the lower bed under the surface of the fluidized bed material. A third tapered screw conveyer fed bed material to the upper bed.

The bed was fluidized by either 100% nitrogen (pyrolysis), 100% air, or mixtures of air and nitrogen. Fluidizing gas passed through an electric heater until bed material reached 300°C. Waste was then used to bring the combustor to 500°C.

The filter system consisted of seven, sintered-metal filter tubes, 2 cm in diameter, in each of five separate filter holders. Each filter holder was piped in parallel so that the exit flow from one holder could be stopped and the flow reversed to remove collected dust while flue gas continued flowing through the other filter tubes.

Air-ejection downstream of the filters provided the motive force for gas flow through the combustor and filter systems. The ejector was adjusted to provide atmospheric (or slightly lower) pressure in the reactor at the point where waste was introduced into the lower chamber. This insured minimal air leakage as feed was added. The air-jet ejector also provided oxygen needed for combustion in the afterburner.

Flue gas passed up from the ejector through a packed bed catalytic afterburner. The afterburner was 61.0 cm in diameter and 91.5 cm long. Additional combustor and cooling room air was pulled in by the negative pressure in the unit and mixed with the process gas stream at three levels in the afterburner.

Flue gas goes through one stage of high efficiency particulate air (HEPA) filtration before leaving the incinerator room. It then passes through four stages of HEPA filtration before venting to the atmosphere.

As in the laboratory scale tests, a fluidized bed of Na_2CO_3 was used to neutralize HCl at its point of generation in order to avoid corrosion and eliminate the need for flue gas scrubbing. Tests were conducted with continuous waste feeding to a single Na_2CO_3 bed. Fresh Na_2CO_3 was continuously fed to the bed and bed material was continuously discharged. Almost 100% efficiency of HCl reaction was obtained up to about 26% Na_2CO_3 utilization. Chloride reaction efficiency decreases and significant quantities of HCl are released after this point (40).

These data were generated with a range of 500-1000 micrometer Na_2CO_3 bed material. X-ray microscopy of particles from used beds indicates that a highly concentrated, dense shell of NaCl surrounded an unreacted Na_2CO_3 core. This indicates that smaller particles would probably improve the neutralization efficiency at higher Na_2CO_3 utilization levels. It also suggests abrasion of the NaCl shell, exposing the Na_2CO_3 core, might also improve Na_2CO_3 utilization.

Neutralization was not affected by the fluidized bed depth. Data from deep and shallow bed experiments indicated that HCl was generated and reacted soon after waste is introduced. This was supported by the fact that the use of a secondary fluidized bed did not improve neutralization efficiency at comparable sodium carbonate utilization levels.

Some catalyst is elutriated from the reactor and from the fluidized bed catalytic afterburner. The rate of loss increases with increased fluidizing air velocity and with small catalyst particles.

Nitrogen was usually mixed with air in the fluidizing medium to prevent overheating. Overheating can cause above bed burning, melt entrained bed material and block the distributor plate. Since additional air is required in the catalyst area to complete combustion of flue gas vapors, the addition of air directly into the catalyst should eliminate this problem (40).

Disposal of Munitions by ARRADCOM/Bench Scale to Pilot Plant Tests

Bench-Scale Tests—

The U.S. Army Armament Research and Development Command (ARRADCOM), Dover, New Jersey, conducted bench-scale tests on the feasibility of fluidized bed combustion of propellant and explosives in a fluidized bed combustor.

The system selected for investigative studies was 0.15 m in diameter and 2.74 m (9 ft) high. It was designed to accept a solid/water slurry feed and had a dry explosive feed rate of 3.18 kg/hr. The bed was sized so that it could be fluidized with approximately 50% of the anticipated requirement of 120% stoichiometric air. This improved the flexibility of the incinerator as it allowed the system to operate in either a one or two stage combustion mode, i.e., all the air could be fed into the bottom part of the bed, or part of the air could be fed into the bottom part of the bed, with the other part of the air fed into the upper portion of the bed. Alumina was used as the bed material. The system also included a slurry feed system, cyclone particulate collector, and stack gas analyzer. Major particulates in the cyclone were alumina fines. The slurry feed system had a mix/feed tank with a large recirculating line. The incinerator feed was tapped from this line and fed into the incinerator through a metering pump (41).

In a series of 37 test runs, made in both the one-stage and two-stage mode for up to 6 hours, the incinerator operated effectively in disposing of the explosives and propellants. However, emission levels of 840 ppm-NO_x, 650 ppm-CO, and 350 ppm-HC (hydrocarbons), were above the 200 ppm emissions goal for each of these pollutants. The emissions were also approximately equal to untreated emissions from previous combustion studies done in rotary kiln and vertical incinerators.

In another series of tests, the addition of 6% by weight of nickel oxide catalyst to the bed caused a reduced emissions to 57 ppm NO_x, 40 ppm-CO, and 10 ppm HC. The results of this program led to the decision to convert an ARRADCOM vertical incinerator to a fluidized bed incinerator for pilot plant testing (41).

Preliminary Pilot-Plant Tests--

Before the pilot-plant ARRADCOM fluidized bed combustion unit began operation, laboratory test runs at Exxon Research Corporation established combustion parameters so that the pilot plant could be operated in an efficient and ecologically sound manner. Optimum combustion occurred if the fluidized bed (with catalyst) was operated with a reducing atmosphere in the lower bed and an oxidizing atmosphere (via secondary air) in the upper bed. The enriched oxygen of the bed, combined with the mixing of the alumina and waste, promotes efficient combustion, thereby minimizing hydrocarbon and CO emissions. Use of a nickel catalyst plus reducing conditions in the lower bed minimize NO_x emissions. The nickel catalyst also promotes the reduction in levels of gaseous pollutants such as CO, and HC. Use of supplemental oil injection assists in maintaining a constant bed temperature and provides a reducing atmosphere in th lower bed. NO_x is formed based on parameters such as combustion temperature, reaction rate, residence time, concentrations of nitrogen and oxygen, and quench rate.

An equivalence ratio was calculated for the first combustion zone (ϕ_1), and for the overall process (ϕ_2). The equivalence ratio is calculated by comparing the actual fuel/air (F/A) ratio to the theoretical ratio for stoichiometric conditions. If the equivalence ratio is 1, the reaction is stoichiometric. Less than 1 equals oxidizing conditions, greater than 1 represents reducing conditions (41).

The fuel/air ratios used, govern operating temperatures required for the material burned. Tests showed that the optimum heating temperature for TNT is 900°C in the slurry ingestion zone. When the system is stabilized at this temperature, the gaseous emissions (especially NO_x) are at their lowest values. For Composition B, the best temperature is 1,038°C (41).

The principle pollutant emissions from combustion of TNT and Composition B were characterized :

- Sulfur oxide emissions, from combustion of fuel oil that contains sulfur, produce significant quantities of sulfur oxides.
- Nitrogen oxides are formed by thermal fixation of atmospheric nitrogen in high temperature processes and from nitrogen compounds in the waste. Because of the relatively high fluidized bed temperatures, only nitrous oxide is formed.
- The amount of CO emitted indicates the efficiency of the combustion process. An efficient combustion operation is associated with high CO_2 and reduced CO levels.
- Particulate emissions from the fluidized bed incinerator can be either dust-solid particles or smoke-solid particles. The dust-solid particles are composed primarily of bed material and catalyst fines entrained in the gas stream. Smoke-solid particles are formed as a result of incomplete combustion of carbonaceous materials. Their diameters range from 0.05-1 micron (41).

Pilot Plant Tests at ARRADCOM--

The pilot plant fluidized bed incinerator was 2.4 in diameter and 9 m tall. The combustor shell was made from 15-cm, Schedule 40, RA-330 high temperature alloy pipe. RA-330 is austenitic, non-hardenable, and strong at high temperatures. It is also resistant to oxidation, corrosion, and carburization. Nominal composition was 19-35-43-1.5-1.25 (Cr-Ni-Fe-Mn-Si). The slurry preparation and feed system can mix and pump various explosive slurries. The entire system was remotely operated and monitored. Slurry was mixed in two, 1.6 m tanks. The tanks were loaded with water and mixers were started. After the water was in motion, the explosive material was added to the tank and mixed with the water. The mixers were operated with a 735 W pneumatic motor. They were adjusted for each individual slurry mix and for liquid level changes. Because of their different compositions, weight percentages, and particle configurations, the densities of the slurries varied and presented different mixing parameters.

The slurry pump was centrifugal and capable of pumping a 18.3 m head with a 0.0079 m^3/sec water flow. Slurry flow through a 0.064 m header pipe was adjusted from the control room (41).

Bed temperatures were carefully monitored to determine where the fuel oil and/or slurry were combusting. Close control was required to prevent combustion from occurring above the bed or in the exhaust ducts. Six slurry injection nozzles were alternated with oil nozzles around the chamber's periphery. Injection of oil and slurry into the bed was controlled by individual "approval switches" set at predetermined combustion temperatures in the bed. This prevents the oil and slurry from being fed into the bed until combustion temperatures are attained. Temperature controls also shut down the system if maximum set temperatures were exceeded. The system proved to be an excellent heat sink. Heat retained after a weekend shutdown was sufficient for a Monday morning startup without use of the preheater.

Pressure was monitored in the plenum, grid, bed, and upper chamber. Plenum and grid pressures indicated degree of fluidization. Grid nozzles could also be checked for clogging. Pressure transmitters could detect pressure buildup in the system and possible detonation. Monitoring of slurry flow was achieved by utilizing pressure transmitters in the header and slurry lines (41).

Alternate Fuel Studies--

Because large incinerators would have rather high pre-heat fuel costs for liquid fuels low in sulfur and nitrogen, in addition to expensive gaseous fuels, studies were made regarding high sulfur and high nitrogen fuels. A 30 wt% nickel catalyst was used. Major results indicated satisfactory operation of a catalytic fluidized bed with both gaseous and low sulfur fuels. Initial tests revealed high sulfur fuels poisoned the nickel catalyst with subsequent high NO_x emissions, although stable combustion was obtained. Further investigation of various parameters is necessary to fully evaluate the relationship of the nickel catalyst with high sulfur fuels (41).

Munitions/Bench Scale to Pilot Plant Processes

Bench Studies at Picatinny Arsenal—

Under the technical direction of the Manufacturing Technology Director-
ate, Picatinny Arsenal, Dover, New Jersey, several incinerator designs were
evaluated to develop a reliable, safe method for waste explosive material
disposal. The fluidized bed incinerator design was judged to be one of the
more promising systems and was selected for further testing (42).

A laboratory scale incinerator was constructed with a 15.42-cm bed
diameter and 3-cm height. Bed material was aluminum oxide granules
(particle size, 500 μ). Waste explosive trinitrotoluene (particle size,
0.318 cm) was introduced to the incinerator bed as a 10 wt% TNT/water
slurry from a mixing tank through a recirculating line. The incinerator
bed material was heated to 871°C-1093°C. The method and rate of air feed
were not indicated, but bed material took on all the properties of a fluid and
provided violent agitation and turbulence. A dry cyclone was placed in the
emission gas stream to remove particulate matter.

During the course of the lab testing, it was observed that a reducing
atmosphere could be created in the bed through use of two stage combustion
and introduction of a nickel oxide catalyst to the bed. This accelerated
the following reaction:

$$2 \ NO + 2 \ CO \ \longrightarrow \ 2 \ CO_2 + N_2$$

Large (unspecified) reductions in NO_x emissions were observed in addition to reductions in CO and hydrocarbons. It was reported that this
eliminated the need for a wet scrubber system and the associated liquid
disposal problem.

Pilot-Plant Combustion at Picatinny Arsenal—

A vertical induced draft incinerator facility, located at Picatinny
Arsenal, was converted into a fluidized bed pilot-plant combustion unit.
It had facilities for slurry preparation and incineration as well as a dry
cyclone separator and a 38-m emission stack. The fluidized bed was
composed of 90% alumina particles and 10% nickel oxide-coated alumina
particles (500 μ or smaller). Nickel oxide was used as a catalyst to
reduce NO emissions.

Waste explosive materials were sprayed into the combustor as a water
slurry at a maximum concentration of 25 wt% solids to water. Emission
streams passed through a dry cyclone separator with a selection efficiency
rated for particles 30μ and above before exiting through the primary stack.
The system was oil fired and designed to operate at bed temperature ranges
of 871-1093°C.

The Air Pollution Engineering Division (APED), United States Army
Environmental Hygiene Agency (USAEHA), conducted an air pollution

assessment program during the initial trial operation of the prototype facility. The objectives were:

1. Monitor air pollution emissions leading to compliance with federal, state, and local regulations
2. Test for hazardous air pollutant emissions from the facility
3. Provide emissions data useful in the evaluation of selected operating parameters and combustion efficiency (42)

Because of the unique character of this prototype incinerator, only New Jersey stationary source emission standards governing visible emissions were found to apply to this facility. An analysis of the prototype incinerator combustion cycle was performed to determine potential direct emissions or by-product formation and emission of hazardous substances.

Several substances introduced into the incinerator were determined to be potentially hazardous air pollutant emitters. These substances include bed material, and the waste explosives TNT and RDX (cyclotrimethylenetrinitramine, $(C_3H_6N_6O_6)$. The conditions in the reactor favored the formation of the toxic compound nickel carbonyl ($[Ni(CO)_4]$). CO reacts with metals to form a carbonyl when a metal, such as the nickel catalyst in the bed, presents a high surface area. Because the formation of carbonyl is favored at low temperature, the greatest potential for its formation is at startup and shutdown. According to a computer simulation program developed in 1971 by the National Aeronautics and Space Administration (NASA) designed to identify the chemical compounds produced by rocket fuel fluidized bed combustion, inorganic cyanides and hydrogen cyanide may be present as contaminants in explosives.

The following pollutants were selected for monitoring:

nickel and alumina particulate matter	HCN
nickel carbonyl	NO
inorganic cyanides	NO_x
TNT	CO
RDX	

No relevant air pollution emission standards were found to apply to these substances. In the absence of legal standards, stack emission limitation criteria were developed by the Air Pollution Engineering Divsion to serve as a reference for decision making during the initial testing period. The procedures for the emission guidelines were derived from the Colorado Air Pollution Commission regulations. They were also based on meteorological dispersion model estimates for the fluidized bed incineration stack, assuming worst cast stability, so that ground level concentrations would not be allowed to exceed 1/30 of the Threshold Limit Value (TLV) established for these substances.

A two week field testing plan was developed which covered a selected range of slurry compositions and percent solids. Variation in incinerator operating temperatures, combustion air flow rates, and fuel input rates were left to the discretion of the operator. All significant operating parameters were recorded during each test (42)

TABLE 10
TOTAL PARTICULATE MATTER AND VISIBLE EMISSIONS DATA

Test Number	Slurry Composition	Particulate Emission Rate Corrected[a] gm/sm³	(gr/scf)[b]	Particulate Mass Emissions[c] kg/hr	(lb/hr)	Visible Emissions[d] % Opacity
1	5% Comp B	1.1245	(0.4914)	1.9450	(4.2880)	4.6
2	15% Comp B	1.1522	(0.5035)	3.1032	(6.8414)	4.8
3	15% Comp B	3.5245	(1.5402)	2.0115	(4.4346)	7.3
4	25% Comp B	0.1483	(0.0648)	0.4455	(0.9822)	5.0
5	5% TNT	2.0758	(0.9071)	2.0740	(4.5723)	10.0
6	15% TNT	1.4524	(0.6347)	2.2401	(4.9386)	16.3
7	25% TNT	2.7931	(1.2206)	2.7647	(6.0952)	11.7

a– Particulate emission rate corrected to 12% CO_2 with the contribution of the auxiliary fuel to CO_2 also discounted.

b– The New Jersey incinerator emission standard for incinerators burning normal type wastes is 0.229 gm/sm³ (0.10 gr/scf).

c– The Colorado process weight particulate emission standard, based on the input of slurry feed, would be 1.05–1.52 kg/hr (2.34–3.36 lb/hr).

d– Opacity values represent the average of the worst 24 consecutive 15 second observations recorded over the monitoring test period.

TABLE 11
GASEOUS AND VAPOR PHASE HAZARDOUS AIR POLLUTANTS EMISSIONS DATA

Hazardous Pollutants Monitored	Maximum Emission Rate Observed (mg/m³)	Emission Limitations Developed by USAEHA (mg/m³)
Nickel Carbonyl	0.028*	210
TNT Explosive	0.006*	890
RDX Explosive	0.010*	890
Inorganic Cyanide	3.400	2970
Hydrogen Cyanide	3.540	6530

* These values represent the sensitivity of the analytical procedures used.

Source: Carroll, J.W. et al., USAEHA, 1979.

Results of the study—Weather conditions were overcast during approximately 50% of the test periods; a condition that hampered evaluation of the gray-white incinerator stack plume. Visible emissions, evaluated as percent plume opacity, were generally found to be in compliance with the 20% limit applicable in New Jersey.

Total munition particulates, expressed in terms of both emission concentrations (gm/sm^3) and mass emission rates were generally found to be high (Table 10). Although no particulate emission standards directly apply to this prototype incinerator, emission levels exceed most of the guidelines of the New Jersey standard for incinerators of this size and the Colorado process weight particulate emission standard. However, the particulate emissions represent a significant improvement in particulate matter emission control versus standard disposal via open burning and detonation of waste munitions. Carroll et al. believe that definition of regulatory emission standards for particulate matter and other potential pollutants for this type of disposal process, in lieu of open burning, should be developed (42).

Monitoring indicated levels for all potentially hazardous vapor and gaseous phase pollutant constituents were found to be negligible (Table 11). All test results for nickel carbonyl and vapor phase TNT/RDX were found to be less than the detectable limit (sensitivity) of the analysis procedures. Trace quantities of total cyanides and hydrogen cyanide were observed on several tests when lower incinerator combustion temperatures were used. These levels were several orders of magnitude less than the emission limit criteria that were developed earlier. Nickel carbonyl was the only constituent found in significant quantities, yet these levels were at least tenfold less than emission limits criteria developed by the USAEHA (42).

HAZARDOUS WASTES DESTROYED BY UV/OZONATION

2,3,7,8-Tetrachlorodibenzo-p-Dioxin

A preliminary investigation of the effect of an UV/ozone system on chlorinated dibenzo-p-dioxins (TCDD) was conducted by California Analytical Laboratories, Inc. and the Carborundum Company (both in Sacramento, California). Chlorodioxins are implied carcinogens and extremely toxic (43).

Experimental Apparatus and Methods—

Two types of ozonation systems were used:
(1) Purified oxygen was passed in front of a mercury vapor lamp. The generated ozone was then bubbled through an aqueous solution of TCDD (1.0 ppb) in a 100 ml volumetric flask. The flow rate, 150 ml/min and ozone content 0.25 mg/l, yielded an ozone dosage of about 0.04 mg/min. After ozonation, 1.0 ml of a mixture of benzene-hexane (1:1) was added to extract any organic compound that might be present. The extract was analyzed by gas chromatographic equipment equipped with an electron capture detector.
(2) The second ozone system was an Ozone Research commercial unit. Generated ozone was passed through a diffuser into the reactor at

a rate of 0.5 l/min. Ozone content in the oxygen was 5 mg/l, and the resulting dosage was 2.5 mg/min. The ozonated solution was then extracted and analyzed as above (43).

Results—

Figure 16 shows the results of the ozonation of a 1 ppb solution of TCDD at pH 3.5 and 10.5. Since the ozone dosage was only about 0.44 mg/min for both pH levels, twelve hours was needed for complete degradation. The pH did not have a significant effect on breakdown time, and the small difference in degradation rates was within experimental error. A 10-15% rate increase was observed when the solutions were both ozonated and irradiated with UV light (300-400 nm) at the same time.

Figure 17 shows results obtained when a 1 ppb TCDD solution was ozonated at pH 7 in the commercial unit. The 2.5 mg/min ozone dosage greatly increased the degradation rate as compared to the previous unit. There were insufficient data to indicate if the addition of UV to the process had a significant effect on the rate.

In order to detect chlorinated degradation products from TCDD resulting from ozone breakdown of TCDD, a Finnigan Model 3200 GC-MS system equipped with an electron capture detector coupled with an Incos computer system was used. To date, no chlorinated compounds have been detected (1979). The more sensitive Multiple Ions Detection mode also failed to detect any chlorinated breakdown products. Detection difficulties could possibly be attributed to the low levels of starting material. Future work will use larger amounts of starting material and apply compound labeling for mass balance studies.

The degradation rate of octachlorodibenzo-p-dioxin (OCDD) by ozonation was also studied. This substance is a common impurity associated with the wood treatment agent pentachlorophenol (PCP). When a 5 ppb OCDD solution (pH 7.1) was ozonated in the commercial reactor at an ozone dosage of 2.5 mg/min, the degradation rate approximated that for TCDD. Approximately 50% was degraded after 40 minutes of ozonation. These preliminary studies indicate the feasibility of ozone or UV/ozone degradation systems for waste chlorodioxins in wastewater (43).

Hydrazine, Monomethyl Hydrazine, and Unsymmetrical Dimethylhydrazine

The hydrazine family of fuels includes hydrazine (H), monomethyl-hydrazine (MMH), and unsymmetrical dimethylhydrazine (UDMH) as well as mixtures of these compounds. A study was conducted for the United States Air Force Engineering Service Center, Tyndall AFB, Florida, by Catalytic, Inc. Philadelphia, Pennsylvania, on the effect of ozonation on these compounds. The effect of ultraviolet light (UV) as a photooxidizer, pH, solution concentration, reactor inlet ozone gas phase concentration, and superficial gas velocity were evaluated. The partial oxidation products of ozone treated hydrazine fuels were characterized and aquatic toxicity testing accomplished (44,45).

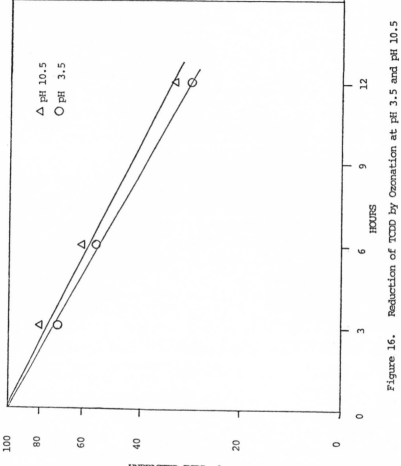

Figure 16.   Reduction of TCDD by Ozonation at pH 3.5 and pH 10.5

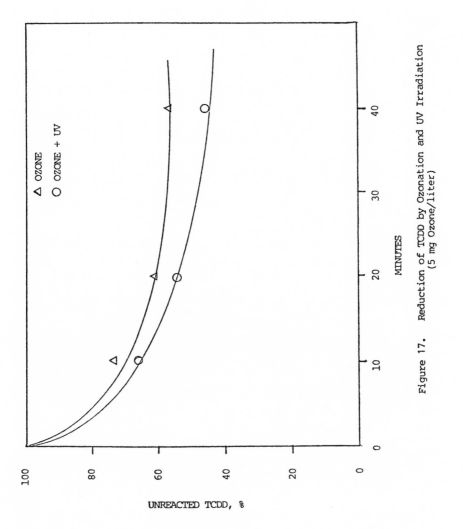

Figure 17.   Reduction of TCDD by Ozonation and UV Irradiation
(5 mg Ozone/liter)

Experimental Apparatus and Methods--

The basic apparatus for all experiments is shown in a process diagram in Figure 18. All experiments were conducted in semi-batch mode (constant liquid supply, continuous gas supply). The reactor was a Life Systems Modified Torricelli Ozone Contactor (LMTOC). A Grace ozonator Model LG-2-L2 produced ozone from both air and oxygen. Air was the ozonator feed gas for runs when an ozone concentration of 13 mg/l (approximately 1% ozone in air) or less was desired. Extra-dry grade oxygen feed gas permitted production of higher ozone concentrations (2% ozone in oxygen) than with air for the same electrical power input to the generator.

Solutions of H, MMH, and UDMH were synthesized from fuel grade material. Thirty liters of material were pumped to the LMTOC from a feed tank for all runs. Samples were extracted at the mid-depth point in the column. Concentrations of H, MMH, and UDMH in the reactor were determined by standard colorimetric methods (44).

All reported trial data were analyzed to determine rate constants (k) for zero, first, and second-order reactions with respect to the hydrazine species involved. In addition, UDMH runs were analyzed for the half-order reaction rate constant (44).

Results--

Only the results of experiments done with MMH were discussed in detail. The authors believed that the findings for H and UDMH generally parallel those found for MMH.

UV photooxidizer—Inlet ozone concentration to the reactor was 10.1 mg O_3/l (with UV light) and 11.3 mg O_3/l (without UV light). Reactor off-gas, measured near the end of both runs, contained 16.7% and 27.8% of the inlet ozone. This difference was due to the presence and absence of UV light, respectively. Figure 19 is a plot of the data from comparative runs in the presence and absence of UV light and shows the positive effect of the UV as a photooxidizer. The zero model predicts reaction half life ($t_{1/2}$-min) values of 19.2 and 26.2 and k (mg/l/min) constants of 3.33 and 2.27 for the experimental run, with and without UV light respectively (44).

pH--The solution characteristics of pH and species concentration are related from a standpoint of chemical kinetics, reaction oxidation pathway, and ozone mass transfer. The pH of the solution greatly affects the rate of oxidation of hydrazine fuels and their partial oxidation products. Previous studies have shown that the oxidation rate of methanol is accelerated at alkaline pH. Solution pH also determines the auto-decomposition rate of dissolved ozone in solution, and therefore, the steady state dissolved ozone residual level that is achievable. Since ozone decomposition leads to oxygen radical production, pH is expected to partially control ozone oxidation of hydrazine fuels. Two pH levels were investigated-highly acidic (pH 2.6) and alkaline (pH 9.1) (44). The effect of solution pH on ozone oxidation of MMH is shown in Figure 20. The graph shows that as solution pH decreased, the oxidation value also decreased.

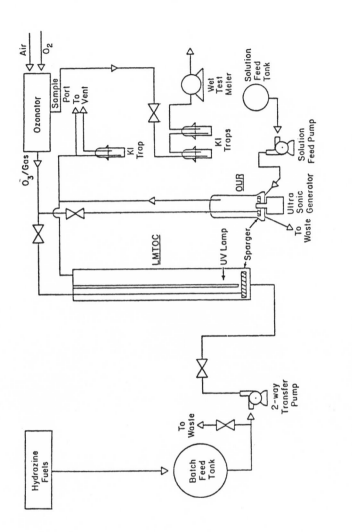

Figure 18.   Process Diagram of Catalytic, Inc. UV/Ozone System

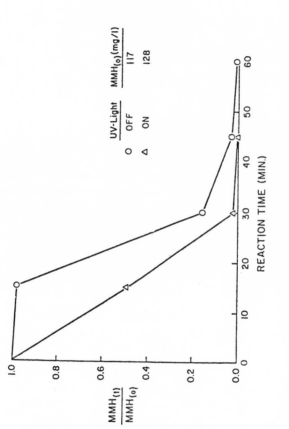

Figure 19.   The Effect of Ultraviolet Light on the Ozone Oxidation of MMH

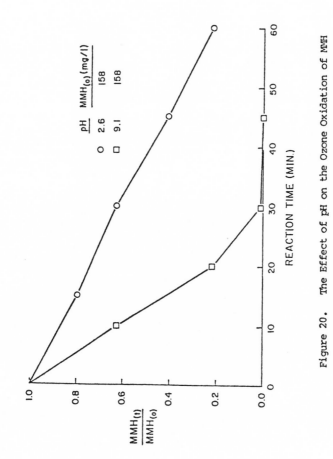

Figure 20.   The Effect of pH on the Ozone Oxidation of MMH

Species concentration effect—In the next series of experiments, ozonator and reactor conditions were held constant, and the disappearance of MMH was followed during ozonation time. Different initial concentrations of MMH (158, 505, and 1171 mg/l) were used. The graph showing the results of this experiment is seen in Figure 21. For the highest MMH concentration run, the reaction appears to proceed through three distinct stages. During the first thirty minutes of ozonation, the reaction is limited by ozone mass transfer. Between thirty-sixty minutes, the dissolved ozone concentration approaches saturation in the LMTOC. Finally, in the last stage, the system appears to be operating under reaction rate conditions. Data from the 505 mg/l run indicates the same response but with a reduced phase I time period. This was expected as the initial mass of MMH present was reduced. As seen by the shape of the curve in Figure 21, the 505 mg/l and 158 mg/l runs did not have mass transfer limitations (44).

The extent of methanol production is related to the MMH concentration remaining in solution and ozonation time. At time zero, the methanol concentration is not zero because of the manner in which the batch was charged into the reactor. The reactor contents were air sparged for eight minutes while pumping MMH from the solution feed tank so that a portion of MMH was converted to methanol. As the initial concentration of MMH increased, alcohol production prior to ozonation also increased (44).

Ozone partial pressure—The amount of inlet ozone gas concentration available to the reactor is related to ozone mass transfer and to the maximum dissolved ozone concentration that can be achieved at fixed reactor operating conditions. Three experiments were carried out with reactor inlet ozone concentrations of 5.2, 10.1 and 27.7 mg O_3/l gas. All other reactor conditions were constant. Ozone concentrations in the reactor off-gas measured 0.66, 1.69, and 2.66 mg O_3/l gas at the end of the low, medium, and high ozone partial pressure run, respectively. Species data for these runs show that MMH disappears readily in all three partial pressure conditions in the LMTOC. The run with an inlet ozone concentration of 5.2 mg O_3/l gas is ozone mass transfer controlled for the first 15 minutes of the reaction (Figure 22). This limitation occurred even though initial MMH concentration was only 89 mg/l (44).

Oxygen sparging—Gas sparging of MMH solutions with air and/or oxygen was studied to find out if this less costly alternative to ozone oxidation would be feasible. These experiments also compared the effect of UV light on this process. MMH solutions were oxygen sparged in the presence and absence of UV light. The experimental results of the runs are presented in Figure 23 (44).

Conclusions—

- The presence of UV light in the reactor reduces $t_{1/2}$ values and increases reaction rates for M, MMH, and UDMH oxidation by ozone over reactions that did not use UV light.
- Increasing solution pH increases the ozone oxidation rate of H, MMH, and UDMH.

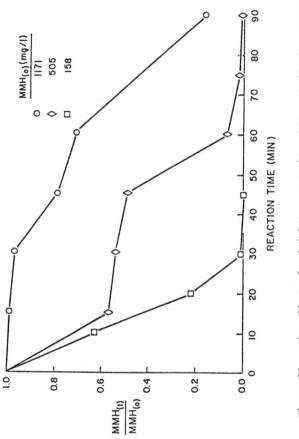

Figure 21.   The Effect of Initial Concentration of Ozone Oxidation of MMH

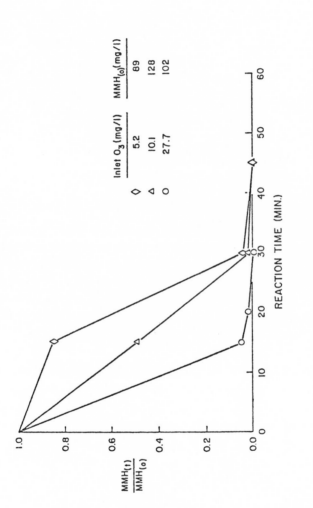

Figure 22.   The Effect of Inlet Ozone Concentration on the Ozone Oxidation of MMH

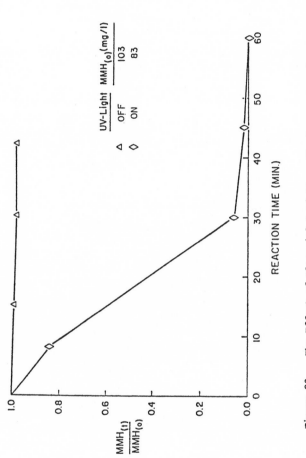

Figure 23.   The Effect of Ultraviolet Light on the Oxygen Sparging of MMH

- Increasing species concentration, at fixed reactor operating conditions (pH, catalyst type, ozone partial pressure, and superficial gas velocity) increases the required hydraulic retention time to achieve the desired effluent concentration for all three hydrazine fuels.
- The quantity of methanol produced from ozone oxidation of MMH is proportional to species concentration.
- Increasing ozone partial pressure decreased $t_{1/2}$ values for H, MMH, and UDMH, but ozone utilization efficiency is reduced.
- MMH decomposition during oxygen or air sparging is greatly enhanced by UV light. The same trend, to a lesser degree, occurs with UDMH and H (44).

Identification of the Partial Oxidation Products of H, MMH, and UDMH—

During the study described previously, attempts were made to identify the partial oxidation products of H, MMH, and UDMH that remained in solution after the fuel was removed by UV/ozonation. The studies involved both chemical and bioassays (44).

Oxidation of hydrazine results in water and nitrogen, according to the following equation:

$$N_2H_4 + O_3 \longrightarrow 2 H_2O + N_2 + O \quad (45)$$

Ozonation produced a small yield of nitrate-N. Ammonia was probably a product of a side reaction and produced nitrate after oxidation.

The main reaction of MMH oxidation would be expected to follow the stoichiometry of the potassium iodate reaction used to determine MMH analytically. This is represented by the following equation:

$$CH_3N_2H_3 + KIO + 2 HCl \longrightarrow KCl + ICl + CH_3OH + N_2 + 2 H_2O \quad (45)$$

This equation predicts that methanol will remain after initial oxidation of MMH is complete. This was verified during the experimental runs. MMH also reacted with ozone to produce nitrate-nitrogen. Although methanol was the only oxidation product identified by from MMH ozonation, at least four other organic compounds were identified from gas chromatography peaks.

Methanol, formaldehyde dimethyhydrazone, formaldehyde monomethyl-hydrazone, N-nitrosodimethylamine, dimethyl formamide, and tetramethyl tetrazene oxidation products were identified in UDMH ozonations. Formal-dehyde and formic acid could also be formed from methanol in both the MMH and UDMH tests.

Since residual organic compounds from MMH and UDMH ozonation are expected to be amenable to ozonation, their concentrations can probably be reduced by continued ozonation past the point of fuel removal. Bio-assay studies indicated ozonation could reduce toxicity levels of methyl hydrazines, however, the resultant wastewater contained some residual toxicity. Consequently, further ozonation is necessary to produce a wastewater safe for reuse (45).

Nitrobenzene

The Westgate Research Corporation of West Los Angeles, California, conducted a study to: (46)
- identify the major oxidizing species in an UV/ozone photooxidized system
- identify the major stable intermediates resulting from UV/ozonation of representative aromatic pollutants
- acquire data to aid in predicting optimum conditions for UV/ozonation of different classes of organic compounds

A study of model compounds and their products in a UV/ozone system would promote the understanding of the nature of the predominant oxidizing species in the UV/ozone system.

Nitrobenzene was selected as the primary model compound based on the ease of gas-liquid chromatographic (glc) separation of the isomeric nitro-phenols, availability of suspected intermediates, and because it is a representative aromatic pollutant. Ratios and amounts of isomeric phenols formed by UV/ozonation of nitrobenzene were compared to the ratios and amounts of the same isomeric phenols formed by exposing nitrobenzene to other, known oxidizing systems. Stable intermediates and pH effects were also studied (46).

Experimental Approach and Results—

Solutions of 0.8 mM nitrobenzene in distilled water were subjected to ozonation and UV/ozonation at a rate of 0.055 mmoles O_3/liter solution/ minute. The ozone generator consisted of a cylindrical stainless steel vessel with an electropolished inside surface. A 40 watt low pressure mercury arc lamp axially located served as the UV light source. Ozone in oxygen at about 2 wt% was introduced at the bottom of the reactor. The solution for ozonation was introduced through an inlet port at the top. Ozone was generated in the laboratory by a silent arc discharge generator. The reactor was operated in the batch mode. Samples were drawn at pre-determined intervals from a valve at the bottom of the reactor. Nitro-benzene solutions prepared in the same manner as those treated with UV/ozone were also oxidized with the following model oxidizing systems Fenton's reagent—a known hydroxyl radical generating system, NaOCl/H_2O_2 which produces singlet oxygen, UV light solely, H_2O_2, and a H_2O_2/UV system in the presence of air or N_2. The amount of H_2O_2 in all these other systems was made equal to the stoichiometric amount of ozone used in ozonation and UV/ozonation.

The oxidized nitrobenzene solutions were extracted with ether/CH_2Cl_2, derivatized with CH_2N_2, and the levels of the isomeric nitrophenols were assayed by glc. Identification was confirmed by GC/MS.

It is apparent that both ozonation and UV/ozonation of nitrobenzene yield very similar ratios of ortho-, meta-, and para- nitrophenols, with the para- isomer predominant. Other oxidizing systems yield different isomer ratios. Both Fenton's reagent and the H_2O_2/UV/air system yield an isomer ratio that is very nearly opposite that obtained by ozone and UV/ozone systems.

Both the ratios and the amount of nitrophenols formed by O_3 and O_3/UV indicated that none of the other oxidizing systems represent the conditions that exist during UV/ozonation and that the conditions in the ozone system in the absence of UV light are very nearly the same as in the presence of UV light. This finding would appear to exclude hydroxyl radicals as the major oxidizing species in the hydroxylation of nitrobenzene by the UV/ozone system.

Similar tests, conducted on benzene, also indicated that the hydroxyl radical or other free radicals are not responsible for oxidations that take place in UV/ozone systems. The authors concluded that ozone itself is the major oxidizing species in these aromatic systems.

Additional information regarding the nature of the predominant oxidizing species in the UV/ozone system was obtained by studying the rates of UV-ozonation of nitrobenzene, benzoic acid, and anisole. The kinetics for these three model compounds and all other aromatic and aliphatic compounds were pseudo first order. A comparison of the rate constants for UV/ozonation of anisole, benzoic acid, and nitrobenzene showed that the reactivities of these compounds were inconsistent with typical UV/ozonation mechanisms, proceeding mainly by electrophilic aromatic substitution mechanisms. This revelation also indicates that hydroxyl radicals do not play a major role in the disappearance of these compounds because the hydroxyl radical behaves as a strongly electrophilic radical in aromatic substitution reactions. For example, the UV/ozonation of anisole would be expected to proceed faster than the UV/ozonation of benzoic acid (46).

Study of Stable Aromatic Intermediates--

During the next phases of the study, aqueous 0.8 mM nitrobenzene solutions were also used, with an ozone input of 20 mg/liter/minute. Data on the nature and change with time of the major stable intermediates formed during UV/ozonation were discussed.

Chromatograms of neutral and derivatized acidic extracts from nitrobenzene solutions that had been ozonated and UV/ozonated were compared. All peaks higher than trace levels found in the chromatogram from UV/ozonation were present and in the same ratio as in the chromatogram from ozonation only. Thus, the hypothesis is reinforced that the oxidation reactions, at least to the point of benzene ring cleavage, are the same for UV/ozonation of aromatics as for ozonation of aromatics. Figure 24 depicts the probable pathway for nitrobenzene ozonation in the presence of UV light. The nitrophenols were the first stable intermediates formed. Simultaneously, benzene ring cleavage without prior hydroxylation is probably also taking place. Dihydroxylation yielded 4-nitrocatechol and two other dihydroxy nitrobenzene isomers which, based on their mass spectra, were probably 3-, and 4-nitroresorcinol. Several minor compounds present had mass spectra that indicted trihydroxynitrobenzene isomers. The major aliphatic intermediates were oxalic acid, formic acid, and glyoxal.

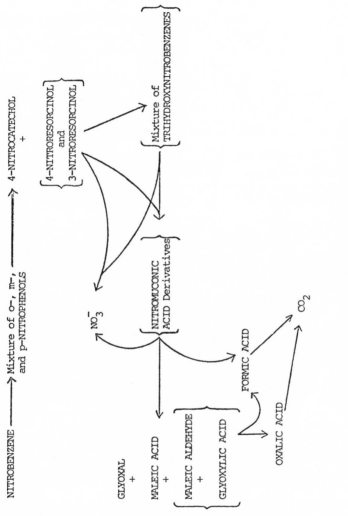

Figure 24.   Nitrobenzene Ozonation Pathway

A small amount of maleic acid was also isolated, and the presence of maleic aldehyde and glyoxylic acid was indicated by formation of their 2,4-dinitrophenylhydrazones.

Similar studies were also determined with anisole, benzoic acid and 3-chlorobenzoic acid as model compounds. The same aliphatic intermediates as in the case of nitrobenzene (formic acid, oxalic acid, and glyoxal) were formed. CO_2 evolution and total organic carbon loss curves were also nearly identical to the previous study.

In a continuance of the study, the major stable intermediates oxalic acid and formic acid were subjected to ozonation and UV/ozonation. UV light increased the rates of ozonation 18 fold for oxalic acid and 6 fold for formic acid (46).

pH Effects—

The effect of pH on the rate of ozonation in the presence and absence of UV light was under study by the authors at the time of publication (1979). Preliminary results indicated that the role of ozone as a nucleophilic reagent is just as important as the usually emphasized electrophilic attack by ozone. Picric acid at a concentration of 0.44 mM was completely ionized and subject to attack by the positively charged oxygen atom of the ozone molecule. In a basic solution, hydoxyl ions may compete with picrate ions for available ozone and thereby inhibit both ozonation and UV/ozonation of picric acid. This competition would not occur in acid solution, and added acid would affect the rate of picric acid degradation very slightly, if at all.

When nitrobenzene is subjected to the same experimental conditions as picric acid, the benzene ring is electron-defecient because of the attached nitro group. The negatively charged oxygen in the ozone molecule may be the primary oxidizing species in the first step of either ozonation or UV/ozonation of nitrobenzene. In acidic solution, hydrogen ions may compete with nitrobenzene for the available ozone. Therefore both ozonation and UV/ozonation of nitrobenzene may be inhibited in acidic solutions.

To summarize this study, UV/ozonation of organic compounds in water at levels of about 0.8 mM is probably not a free radical phenomenon. Major intermediates for the UV/ozonation of monosubstituted benzene derivatives were oxalic and formic acids. In the absence of UV light these acids are relatively stable, but in contrast they are readily ozidized to CO_2 by the UV/ozonation system (46).

Copper Process Waste Streams

System Products Division of IBM Corporation, Endicott, New York, treated two copper removal process waste streams with ozone and ultraviolet light. The waste streams were from a chemical copper recovery process and an electrolytic copper recovery (deplating) process. They contained Na_2SO_4 formic acid, EDTA (ethylenediamine tetraacetic acid), and anionic

surfactant. In order to discharge the treated effluent into the plant clar-
ification system, the final concentration of EDTA should not exceed 5 mg/l.

Experimental Apparatus, Methods, and Results--

   Bench scale tests--An Ultrox Irradiation Model B-803, bench-scale
UV/ozone system was used for all test runs. This system is made by
Westgate Research Corporation, West Los Angeles, California. Ozone was
generated by feeding either industrial grade cylinder oxygen or oxygen from a
cryogenic storage container to an OREC Model 03B2-O ozonator made by Ozone
Research and Equipment Corporation, Phoenix, AZ. The measurement and
control of solution pH was performed with the aid of a Corning Model 12
pH meter that was checked against appropriate buffers for the desired pH.

   In all bench-scale tests, nine liters of waste solution were treated
with ultraviolet light and ozone in the reactor. After the solution was
pumped into the reactor, pH was adjusted between 4 and 6, oxygen gas was
introduced into the system, and the flow rate was adjusted to the minimum
bubbling rate. Bubble rate and size were observed through glass viewing
ports.

   After the ozone generator and UV lamps were switched on, a slow stream
of ozone-rich bubbles moved upward through the waste solution. During this
initial period in the run, the surfactant in the waste was oxidized to the
point where it no longer caused excessive foaming. At this point, oxygen
flow was increased to approximately 192 l/min and ozone generator power was
set to yield 907-998 grams of ozone per day (47).

   All test conditions were the same except that the waste solution from
the deplater contained 16 ppm of iron and 18 ppm of copper, while the
solution from chemical copper recovery contained negligible iron and
copper by comparison. The initial EDTA concentration of both solutions was
approximately 1000 ppm.

   After eight hours, the chemical copper recovery solution had an EDTA
concentration of about 5 ppm. After 12 hours, the EDTA concentration in
the deplater solution was 430 ppm. The deplater solution had become turbid
from formation of extremely fine particles of reddish brown $Fe(OH)_3$.
Although the total amount of iron present in the deplater solution was
small, the effect on the reaction was very significant. If the initial
copper concentration was greater than 30 ppm $Cu(II)$, a large number of
black CuO particles formed that also adversely affected the reaction (47).

Pilot-Scale Tests--

   Pilot-scale test runs were performed using an Ultrox water purification
system, Model P7708 STAC manufactured by Westgate Research Corporation. A
synthetic waste solution composed of reagent grade sodium formate, disodium
ethylenediaminetetraacetic acid, sodium sulfate, 50% sulfuric acid, and
deionized water was used to simulate the chemically copper-recovered addi-
tive waste solution. The solution contained 22 gm/l of formic acid,

100 gm/l sodium sulfate, and 530 mg/l of EDTA (acid form). The quantity of sulfuric acid present yielded a solution at pH 3-4. The solution also contained approximately 2 mg/l of Fe(III) and no detectable Cu(II). The pH was monitored continuously and maintained in the 4-6 range by addition of 50% sulfuric acid as the oxidation progressed. The solution temperature increased from 25 to 45°C during the course of the run due to heat of reaction and heat from the ultraviolet lights (47).

Complete destruction of EDTA and formic acid was achieved for all runs in approximately eight hours using an ozone mass flow rate of 5.5 gm/min at 4% by weight of oxygen. Ozone concentrations were determined iodometrically with a PCI Ozone Corporation ozone monitor. Eight, 65-watt ultraviolet lights were on during test runs.

To verify adverse effects of high Cu(II) and Fe(III) concentrations on the UV/ozone process, deplater solutions containing 16-20 mg/l Cu(II) and 23-28 mg/l Fe(III) were subjected to the process. The initial EDTA concentration was 1677 mg/l with lesser amounts of wetting agent, formic acid, formaldehyde, and methyl alcohol.

After 18 hours treatment with ultraviolet light and ozone (same reaction conditions as used for the synthetic waste), the EDTA concentration was 150 mg/l. Even though a much longer treatment time was used because of the higher initial EDTA concentration, it was estimated that the time required to reach an EDTA concentration of 5 mg/l would easily extend beyond 24 hours. Prolonged treatment time was also observed in bench scale tests of deplater solution where the initial EDTA concentration was lower and the Cu(II) and Fe (III) concentrations were the same or higher (47).

The increase in treatment time for deplater solution is caused by the presence of Cu(II), since Cu(II) can be converted by ozone to minute, black CuO particles. CuO particles, or other copper compounds present in the waste solution, catalyze ozone decomposition to oxygen. This results in destruction of large quantities of ozone, and there is much less ozone available to oxidize organic compounds. Copper concentrations of one or two mg/l are effective in slowing oxidation. Higher concentrations (nearly 50 mg/l) extend treatment time significantly and result in waste solutions that appear totally black (47).

Results--

If oxidation is performed for an extended time period, the organic compounds in waste solutions from electroless copper plating can be completely oxidized using an ultraviolet light/ozone process. Treatment time can be greatly reduced if the following factors are controlled:
- Waste solution pH is maintained in the 4-6 range.
- The Fe(III) concentration is less than 5 mg/l.
- The Cu(II) concentration is less than 1 mg/l.

Maintaining the indicated pH range assures that gaseous carbon dioxide is formed and sparged out of the reactor. This prevents accumulation of

carbonates in the waste solution. In addition, turbidity formation from
$Fe(OH)_3$ is restricted. Low iron concentrations also limit formation of
ferric-EDTA complexes (more difficult to oxidize that the acid form of
EDTA).

Concentrations of Cu(II) higher than one mg/l cause formation of CuO
particles that catalyze ozone decomposition. Sodium sulfate in the waste
solution is not affected by ultraviolet light/ozone treatment (47).

PCB's

General Electric Company Pilot Plant Study--

A pilot plant was set up at General Electric Company's Capacitor
Products Department Facilities in Hudson Falls, New York, to demonstrate
efficiency and cost-effectiveness of the ULTROX UV/ozone system to destroy
PCB's in industrial effluent. Although PCB's have not been used in the
Hudson Falls/ Fort Edward facilities since July, 1977, residual PCB concen-
trations in the untreated effluent ranged from 5-40 µg/l (48).

Experimental apparatus and methods—The portable, skid-mounted pilot
plant was 28" wide, 45' long, and 45'high with a 75 gallon wet volume.
It was fabricated from 304 stainless steel, and passivated plus electropolished to reduce chemical attack and increase UV reflectivity. The
reactor could accomodate up to thirty, 40 watt G36T6L, low pressure UV lamps.

Ozone was produced from liquid oxygen via an ozone generator and
diffused from the base of the reactor through porous ceramic spargers.
Figure 25 shows the pattern of water flow through the reactor. Water
passes through each of the stages in a tortuous path to achieve a greater
degree of plug-flow. In each stage, water is contacted by ozone gas and
UV light (48).

Thirty-seven tests were run and the following variables were investigated UV lamp patterns, influent flow rate, ozone mass flow, ozone concentration, ozone mass flow distribution, pH of influent, and temperature.
Data from these tests are tabulated in Appendix D. From the results of
these tests it appears that most of the test runs had more than one
variable, and the authors drew no specific conclusions concerning the most
favorable parameters for PCB destruction. However, many of the runs
contained no µg/l of PCB in the effluent after treatment, and no run had
more than 4.2 µg/l of PCB. The latter run was the only one which did not
expose the waste to UV light.

Results--

When the data in Appendix D were subjected to computer analysis, a
mathematical model was formulated to simulate the performance in destroying
PCB's both in the ULTROX pilot plant and in full-scale equipment (Appendix
E). Upon completion of a satisfactory model, feed and effluent PCB con-
centration limits were established and capital plus O&M costs were derived.
These costs will be discussed in detail in the section on economics.

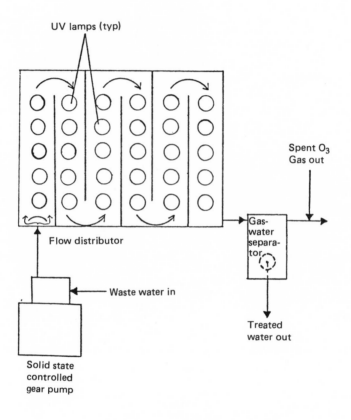

Figure 25.   Schematic of Top View of ULTROX Pilot Plant
(Ozone Sparging System Omitted)

-6-

Hazardous Waste Destruction Technologies
in the Development Stage

BACKGROUND

This section reviews technologies for the treatment of hazardous wastes that are still in the developmental stage. Most of these technologies have been studied only at the bench level; a few are at the pilot plant stage. The technologies reviewed include catalyzed wet oxidation of toxic chemicals, the dehalogenation of compounds by treatment with ultraviolet light and hydrogen, electron irradiation of toxic compounds in aqueous solution, UV/chlorinolysis of hydrazine in aqueous solution, and the catalytic hydrogenation dechlorination of polychlorinated biphenyls (PCB's).

Wet air oxidation experiments were conducted in a titanium autoclave, but only batch oxidations were investigated. The ultraviolet light/ hydrogen technology for the dehalogenation of compounds has advanced from bench scale to the pilot plant stage. This system is not designed to process waste streams with much more than 1% toxic organic content. The electron treatment of trace organic compounds in aqueous solution has been investigated in the laboratory at MIT. The UV/chlorinolysis of hydrazine in aqueous solution has been tested on 8,000 liter batches of wastewater. The catalytic hydrogenation-dechlorination studies of PCB's took place in an autoclave.

A description of each process, experimental details, plus results and discussion follow. The following lists the various chemicals treated by specific processes:

LIST OF TOXIC SUBSTANCES TREATED BY CHEMICAL TECHNOLOGIES

Wet Air Oxidation
2,4-D
Glycolic Acid
Pentachlorophenol
Ethylene dibromide
Malathion
Acetic Acid
PCB's
TCDD (tetrachloro-p-dioxin)
Kepone

Dehalogenation/UV-hydrogen (H_2)
Arochlor 1254 (PCB)
Tetrabromophthalic anhydride
Kepone

Electron Irradiation
2,3,4'-Trichlorobiphenyl
4 Monochlorobiphenyl
Monuron

UV/chlorinolysis
Hydrazine
Monomethylhydrazine
Dimethylnitrosamine
Unsymmetrical dimethylhydrazine

Catalytic Hydrogenation-Dechlorination
PCB's (Arochlor 1242, KC-400

DESTRUCTION OF TOXIC CHEMICALS BY CATALYZED WET OXIDATION

Wet air oxidation (WAO) is a commercially proven technology for the destruction of organics in wastewater and sludges. In conventional wet air oxidation, waste is pumped into the system by a high-pressure pump and mixed with air from an air compressor. The waste is passed through a heat exchanger and then into a reactor where atmospheric oxygen reacts with the organic matter in the waste. The oxidation is accompanied by a temperature rise. The gas and liquid phases are separated, and the liquid is circulated through the heat exchanger before discharge. Gas and liquid are both exhausted through control valves. System pressure is controlled to maintain the reaction temperature as changes occur in feed characteristics (i.e., organic content, heat value, temperature). The mass of water in the system serves as a heat sink to prevent a runaway reaction that might be caused by a high influx of concentrated organics (53).

IT Enviroscience, Inc., of Knoxville, Tennessee, has developed a proprietary catalyzed wet oxidation process based on information in U.S. Patent 3,984,311 (54) (Originally assigned to the Dow Chemical Company, and now assigned to IT Enviroscience for development and commercialization). This process uses a cocatalyst system consisting of bromide and nitrate anions in an acidic, aqueous solution to destroy either organically contaminated aqueous waste or organic residues. Destruction of waste organics is accomplished by mixing a waste with the catalyst system and oxygen (or air) at temperatures greater than 100°C (55).

Experimental Methods

Experiments were conducted in a 1 liter stirred titanium autoclave and limited to batch oxidation. Aqueous wastes or organic residues were pumped into a continuously stirred tank reactor (CSTR) containing a solution of HBr and nitric acid (Figure 26). Air is sparged in, and the organics are

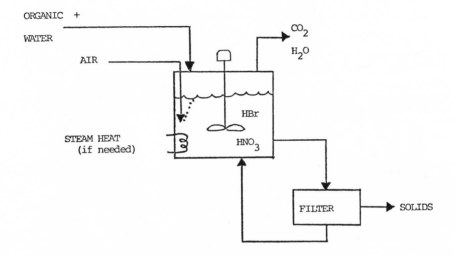

Figure 26.   IT Enviroscience Process Concept
for Homogeneous Catalyst

oxidized with the heat of reaction driving off water. Any solids formed have to be removed, but the catalyst solution remains in the reactor. Carbon dioxide, water vapor, excess air, and any volatile organics formed also leave the reactor. The most important concepts in the process are that non-volatile organics remain in the reactor until oxidized, and there is no bottoms product. Thus, very high destruction efficiencies and low reactor effluent concentrations are not required in reactor design. If the organics remain in the reactor long enough, they will ultimately be destroyed (55).

Measurement of Destruction Rates--

Organic destruction rates were measured by different procedures depending on the water solubility of the organic. The oxidation rate of water soluble organics was measured by withdrawing liquid samples during the reaction and analyzing the organic concentration by Total Organic Carbon (TOC). The destruction rate for insoluble organics was measured by terminating the reaction, cooling the system, and solvent extracting the reactor system and catalyst mixture. The solvent was then analyzed by gas chromatography for unreacted organics and by-products.

The extraction procedure limited the amount of data collected during the experiment, but it was reliable and reproducible. Secondary measurements of the organic destruction were made by inorganic chloride analysis for destruction of chlorinated organics and by carbon dioxide analyses of the final reaction gas. These measurements were used to determine the completeness of the organic destruction. Over 200 runs were made to screen the process efficiency on a variety of organic compounds and measure the effects of different catalyst combinations (55).

Economics

Based on the destruction rates of the organics tested and the process designs described above, preliminary capital and treatment costs have been estimated (Table 14). These estimates show the costs for fast and moderate destruction rates of aqueous wastes (pentachlorophenol or glycolic acid) and the cost for a moderate destruction rate organic residue (Arochlor 1254). The capital cost is the total installed cost using titanium as the construction material (55).

Results

The organics studied can be grouped into two categories, those which oxidize rapidly at low temperatures and those with slower destruction rates that require higher reaction temperatures. Some of the results of the fast destruction rate organics are shown in Table 12. These experiments were conducted at 165°C with a catalyst mixture of 0.5-5% Bromide and 0.1-5% Nitrate. The organic destruction rates, as measured by the disappearance of the initial organic, are much higher than the reported destruction rates of conventional wet oxidation at this temperature. The complete oxidation to carbon dioxide of these compounds was measured by the disappearance of total organic carbon for 2,4-D and glycolic acid and by final carbon dioxide concentration for pentachorophenol, ethylene bromide, and malathion. The

TABLE 12
ORGANICS WITH FAST DESTRUCTION RATES

	Organic Reduction	Total Organic Destroyed	Temperature	Reaction Time In Minutes
2,4-D	65%	65%	165°C	15
Glycolic Acid	99%	99%	165°C	30
Pentachlorophenol	99%	75%	165°C	30
Ethylene Dibromide	94%	51%	165°C	60
Malathion	99%	58%	165°C	60

TABLE 13
ORGANICS WITH MODERATE DESTRUCTION RATES

	Organic Reduction	Temperature	Reaction Time In Hours
Acetic Acid	36%	165°C	1-1.5
Arochlor 1016	93%	195°C	1
Arochlor 1254	95%	250°C	2
Dioxin	99%	200°C	4
Kepone	93%	250°C	6

TABLE 14
PRELIMINARY CAPITAL AND TREATMENT COSTS

	Aqueous (*) Waste		Organic (#) Residue
	Fast Destruction Rate	Moderate Destruction Rate	Slow Destruction Rate
Capital Cost 1979 $	$844,000	$1,260,000	$890,000
Treatment Cost ¢/kg organic $/4000 liters	10.4 21	13.2 27	100

(*) 100 liters/minute, 5% organic concentration
(#) 22.5 kg/hour organic residue

Source: Miller, R.A. et al., IT Enviroscience, Inc., 1980.

total organic destruction rate was 90% for these compounds when the reaction temperature was raised to 200°C (55).

Organic compounds with either low solubility or chemically stable structures are more difficult to oxidize. These compounds include acetic acid, polychlorinated biphenyls (PCB's), tetrachlorodibenzo-p-dioxin, (TCDD), and Kepone. Temperatures up to 250°C were required to destroy these compounds (Table 13). The capability to measure total destruction of the organic by carbon dioxide was not available at the time these experiments were conducted, but gas chromatography of the final reactor contents indicated total destruction of most compounds (55).

Process Design Concepts

Aqueous Organic Waste--

The process design for destroying both aqueous waste and organic residues centers on utilizing the homogeneous catalysts in a continuously stirred tank reactor. The two variations on the basic reactor concept, one for aqueous waste and one for organic residues, differ in the amount of water processed. For dilute aqueous wastes, the energy released by the oxidation of the organics is insufficient to remove all of the incoming and formed water. Therefore, it is necessary to recover the catalyst solution for reuse. Figure 27 shows the basic recovery concept whereby the catalyst could be recovered by evaporation/concentration. The evaporative recovery process requires vaporization of all water entering the system. The water vapor is condensed and discharged separately from the off-gas. Auxiliary heat must be supplied for streams containing less than about 4% organics (depending on the heat content of the waste), and evaporators must normally be used to supply sufficient heat transfer area.

The number of evaporators is an economic decision based on steam economy versus capital. This type of economic evaluation was done for a model plant designed to treat 80 l/min, with the result that a single evaporator offered the best compromise between unit costs, capital, and versatility. For wastes containing high organic levels, no evaporator capacity would be required (55).

Non-Aqueous Organic Waste--

The continuous process concept for treating non-aqueous wastes is substantially different from the process concepts described for aqueous organic wastes since the quantity of water which must be removed from the process is very low. Figure 28 shows the process concept. In many cases, only the water formed as a by-product from the oxidation reaction (plus any amount of water entering the compressed air) must be considered. The only stream normally leaving the process is the off-gas containing principally nitrogen, unused oxygen, carbon dioxide, low levels of water vapor, traces of volatile inorganic (HCl), and organic species which could be present in the reactor mixture.

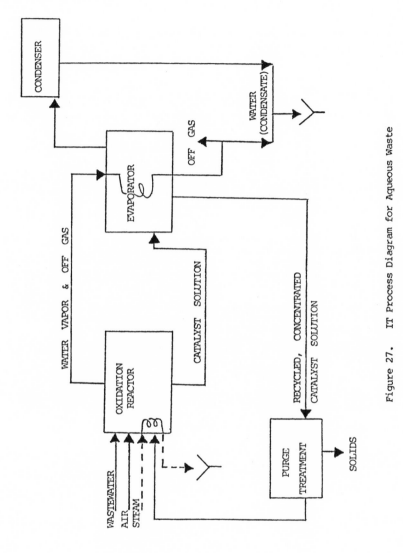

Figure 27.   IT Process Diagram for Aqueous Waste

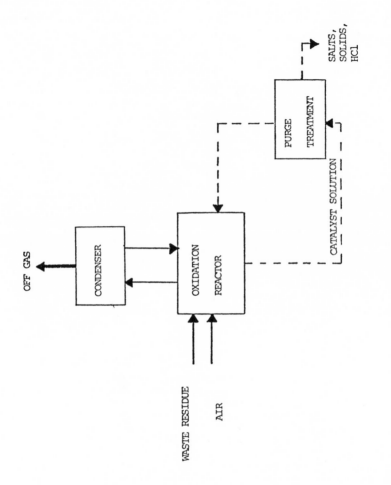

Figure 28.   Process Diagram for Organic Residues

Heat generated from the oxidation must be removed by condensing and refluxing water vapor leaving the reactor. This heat could be recovered by operating the reflux condenser as a steam generator. The catalyst is contained in the reactor and extra unit operations (such as evaporation) for catalyst recovery are not required (55).

Summary

Advantages of the IT Environscience catalyzed wet oxidation process are best understood relative to the conventional technologies of uncatalyzed wet oxidation and incineration. In comparison to straight wet oxidation, the catalyzed process achieves high levels of destruction of a variety of organic chemicals at significantly lower temperatures and pressures. Conventional wet oxidation requires temperatures approaching 300°C and high pressures to achieve greater than 90% destruction of soluble organics. The catalyzed process operates at less than severe conditions. It also produces no aqueous bottoms product; all nonvolatile organics stay in the reactor system until oxidized. The homogeneous cocatalyst enables the system to treat water insoluble compounds.

In comparison to incineration of hazardous wastes or aqueous wastes, the catalyzed wet oxidation process has several advantages. Little or no added energy is required and auxiliary fuel is usually not consumed. It has few unit operations and functions at low temperatures and pressures. Vent gas volume and vent gas scrubber effluent are lower than those produced by incineration and are readily adaptable to polishing treatment if required for control of trace toxic releases. With few unit operations and low volume streams, the oxidation system is potentially portable and can be relatively easily developed in pilot-plant tests (55).

The system is projected to be capital intensive from a cost standpoint. Depending on the type of waste and desired destruction rate, capital outlays of $850,000-$1,300,000 are projected (1980).

DEHALOGENATION OF COMPOUNDS BY TREATMENT WITH ULTRAVIOLET LIGHT AND HYDROGEN

A patent for a process to dehalogenate compounds by treatment with ultraviolet (UV) light and hydrogen (H_2) has been assigned to Atlantic Research Corporation, Alexandria, VA (56). The process is effective for halogenated organic compounds with at least one C-halogen group and works in the absence of any substantial amount of oxidizing agent. The halogenated compound is reduced when carbon-halogen linkages are broken (halogens are liberated during the process). The treatment may result in further degradation of the partially dehalogenated compound. Different degrees of dehalogenation are primarily related to the energy in the C-halogen bond and can be compensated by employing higher or lower energy UV radiation.

Experimental Methods

Figure 29 shows a schematic drawing of the 1.5 liter reactor used in the bench scale process. The reactor is equipped with a 254 nm UV immersion tube, hydrogen gas bubbler, and a recirculation pump. UV radiation

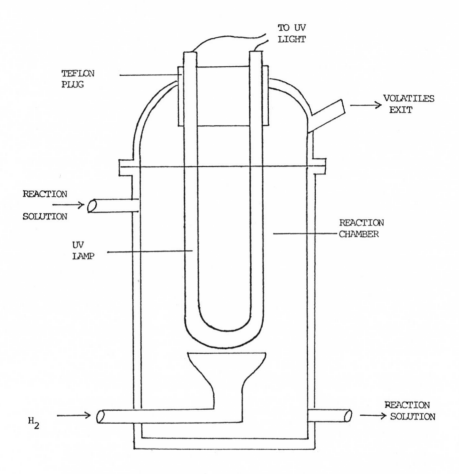

Figure 29.   Schematic of Atlantic Research 1.5 liter Reactor

for these experiments was at 254 nm, since it has been established there is high absorptivity by halogenated compounds at this wavelength. The UV tube is positioned longitudinally in the reactor chamber and held in an air-tight position by teflon plugs. The UV tube is connected by wires to a transformer (not shown). Hydrogen gas is pumped in via an inlet tube. Reaction solution is pumped in via another inlet tube and continuously recirculated by pumping through an outlet tube (not shown). Another outlet tube provides for the exit of volatiles (56,57).

The process can be employed for compounds in the gaseous, liquid, or solid state. Halogenated compounds in the liquid or solid form should be in a finely divided form and dissolved in a suitable solvent which is substantially transparent to the UV wavelengths used. The type of solvent used should be determined by the solubility characteristics of the compound of interest. Alkaline solutions of water, methanol, ethanol, 1- and 2-propanol, hexane, cyclohexane, and acetonitrile are examples of solvents used in the process. Alkalinity is preferably produced by the presence of alkali metal oxide or hydroxide to minimize potentially destructive anions. An organic compound such as methanol is used to solubilize compounds insoluble in water (56).

Reaction Mechanisms

Possible reaction mechanisms suggested by the authors include:
- formation of an excited molecule followed by homolytic dissociation of carbon-chlorine bonds and hydrogen abstraction by the radicals produced from both hydrogen gas and water molecules
- a bi-molecular reaction of the excited species with hydrogen gas or water

Results and Discussion

The process has been demonstrated successfully on Arochlor 1254 (a polychlorinated biphenyl) in methanol and on tetrabromopthalic anhydride in methanol. However, the majority of work has been done on Kepone, both in methanol and water. Degradation of Kepone was found to be pH dependent. When reactions were run in 0.1, 1.0, and 5% NaOH, the latter caustic level gave the best degree of degradation. More than 99% removal of Kepone has been observed in less than 90 minutes. In tests that compared the UV + H_2 process with UV/ozone or straight UV treatment, the quantity of unreacted Kepone or Arochlor 1254 remaining after a given time was always less for the UV + H_2 runs.

Several degradation products were observed including mono-, di-, tri-, tetra-, and pentahydro derivatives of Kepone. Isotopic studies performed indicated that the hydrogen in these compounds comes from hydrogen added during the process.

Based on successful bench studies, a pilot plant for the treatment of higher levels of Kepone was built by the Atlantic Research Corporation and was in use during 1980. Economic data regarding the system is currently

unavailable. The successful treatment of Kepone implies the system should
be adaptable to PCB's and a variety of other organic hazardous wastes in
the ppm or lower range (56,57).

ELECTRON TREATMENT OF TRACE TOXIC ORGANIC COMPOUNDS IN AQUEOUS SOLUTION

As part of a National Science Foundation-sponsored program at the
Massachusetts Institute of Technology (MIT), the effect of electron bombardment on trace toxic organic compounds was studied in pure water solutions
and in model systems containing trace organics in water. Trace organic
compounds studied included Monuron (a persistent herbicide of the urea
type), 2,3,4' trichlorobiphenyl, and 4 monochlorobiphenyl (58).

Experimental Methods and Apparatus

Samples were treated with 3 million volt electrons from the Van
de Graaff accelerator at MIT's High Voltage Research Laboratory. Pre- and
post- irradiation samples were analyzed using a Waters #204 reverse
gradient high pressure liquid chromatograph (HPLC). The reverse gradient
indicated both the "parent" compound and the creation of molecular frag-
ments. All samples were detected by UV absorbance at 254 nm; peak heights
of the parent compounds were found to be linear with concentration.

Master solutions were made by adding a concentrated solution of the
test material to a 1 liter bottle of acetonitrile, allowing the acetonitrile to evaporate, and adding 1 liter of twice-filtered water. The
resulting solution was allowed to stand for about a week to come to
equilibrium. For irradiation, 2 ml of the master solution were placed in
closed glass Petri dishes with a liquid capacity of 6 ml. The Petri dishes
and their contents were passed through the electron beam on a moving belt.
Radiation dose was controlled by adjustment of the product of electron beam
current and exposure time (58).

Monuron

Figure 30 shows the chromatogram of Monuron (0.4 mg/l in water) by
reverse phase gradient elution HPLC before irradiation. Figures 31 and
32 show the results of 10 and 100 kilorads irradiation respectively.
In Figure 31, a vestige of the original parent peak remains with three new
peaks (representing degradation compounds) to the left of the parent peak.
These degradation products are more soluble in water than the precursor.
After 100 kilorads irradiation (Figure 32), the original peak and neighbor
degradation peaks are completely removed. To the far left, there is a
broad peak of highly water-soluble residue. Figure 33 summarizes the data
concerning percent degradation versus radiation dosage. Substantially
complete destruction is attained at 30 kilorads (58).

2,3,4'-Trichlorobiphenyl

2,3,4'-Trichlorobiphenyl has low water solubility. When dissolved in
water to the saturation limit (80 ppm), it was totally destroyed by all
radiation doses down to 1 kilorad (58).

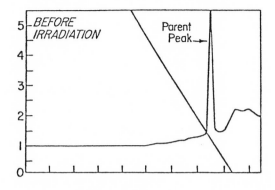

Figure 30. Monuron in Water at 0.4 mg/l, Standard before Irradiation.

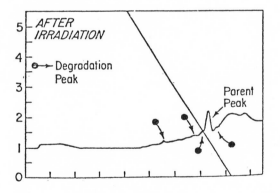

Figure 31. Monuron in Water at 0.4 mg/l, Exposed to 10 Kilorads.

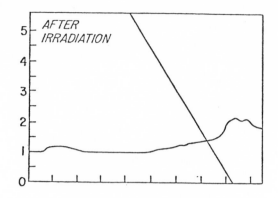

Figure 32.   Monuron in Water at 0.4 mg/l, Exposed to 100 Kilorads.

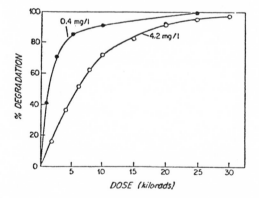

Figure 33.   % Degradation Vs. Dose for Monuron in Water, 0.4 mg/l and 4.2 mg/l.

4-Monochlorobiphenyl

4-Monochlorobiphenyl was somewhat more resistant to degradation than 2,3,4'-trichlorobiphenyl and less resistant than Monuron. When dissolved in pure water to a concentration of 0.8 mg/l, a dose of 10 kilorads was required to achieve nearly complete degradation (58).

Summary and Discussion

Although the Van De Graaff generator successfully treated toxic organics, scale-up of the system would be projected as highly cost-intensive. Few hazardous waste generators have access to a high energy accelerator. Since the test samples were placed in Petri dishes and irradiated on a moving belt, little can be inferred as to the type and cost of scale-up feed mechanisms. Moreover the organics were irradiated in nearly pure water. The attenuating effect of turbidity may be substantial when irradiating a heterogeneous water system.

Consequently this technology, although successful under ideal laboratory conditions, does not appear to be amenable to scale-up from either a cost or industrial standpoint.

UV/CHLORINOLYSIS OF HYDRAZINE IN DILUTE AQUEOUS SOLUTION

A process was developed at Rocky Mountain Arsenal for the United States Air Force by the IIT Research Institute for the UV/chlorinolysis treatment of wastewater containing hydrazine (HZ) in concentrations varying from a few to several thousand parts per million. Some of the wastewater also contained varying amounts of monomethylhydrazine (MMH), dimethylnitrosamine (DMNA), and unsymmetrical dimethylhydrazine (UDMH) (59).

Chlorinolysis Reactions

Chlorine reacts with the contaminants by the following reactions:

(HZ) $N_2H_4 + 2\ Cl_2 + H_2O \longrightarrow 4\ Cl^- + N_2 + H_2O$

(MMH) $(CH_3)HNNH_2 + 4\ Cl^- + H_2O \longrightarrow CH_3OH + N_2 + 4\ HCl$

(UDMH) $(CH_3)_2NNH_2 + 4\ Cl^- + 2\ H_2O \longrightarrow 2\ CH_3OH + N_2 + 4\ HCl$

(DMNA) $(CH_3)_2N_2O + Cl_2 + 2\ H_2O \longrightarrow 2\ CH_3OH + N_2 + HOCl + HCl$

Thus, two moles of Cl_2 are required for each mole of hydrazine and one mole of Cl_2 is required for each mole of DMNA (59).

Experimental Methods and Apparatus

After initial testing on the bench scale, a plant was constructed that could process wastewater in 8,000 liter batches. The flow sheet for the process is shown in Figure 34.

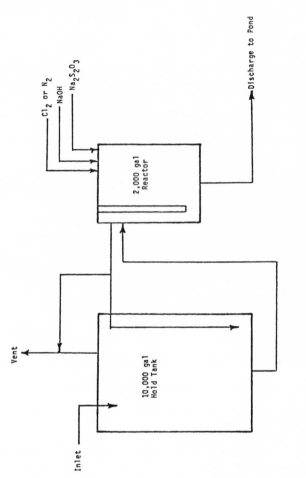

Figure 34. Process Flow Diagram UV Chlorinolysis Reaction System

A 40,000 liter hold tank is used to collect the wastewater and to level out variations in the concentration for more consistent day-to-day operation of the facility. The chlorinolysis reactor is a 8,000 liter glass lined reactor vessel. It is equipped with an agitator and a sparger to insure good reactant mixing and contact. A UV light is immersed in the reactor to activate the chlorinolysis reaction. A total of 7,500 watts of low pressure Hg lamp UV are required. Runs were done both with the UV lamp on and off.

The pH and chlorine concentrations of the reactor contents are monitored and controlled. The pH is maintained by the addition of 50 percent NaOH. Chlorine is supplied as a gas to the reactor. The chlorine addition system is controlled by an oxidation/reduction potential instrument (ORP) located in the reactor recirculation loop. Control valves are installed in the supply line to turn the chlorine flow on and off as required. The valves, piping, and sparger are the major components of the system and are constructed of stainless steel.

After the reaction is completed, the excess chlorine is stripped by nitrogen to less than 100 ppm. Residual chlorine is then neutralized with $Na_2S_2O_3$, the pH is adjusted to 7.0, and the reactor contents are discharged to a holding pond. The effluent may then go to a biological treatment facility, to land spreading, or to a waterway (59).

Results and Discussion

Samples of the end products of the chlorinolysis process with and without UV were analyzed for end products. Even though the initial end products of the chlorinolysis reactions (N_2, CH_3OH) are not especially toxic, they may further react with chlorine to form compounds such as CH_3Cl, CCl_4, and nitrogen trichloride (NCl_3). NCl_3 is the most undesirable of these since it is explosive.

When samples of the end products of the process for chlorinolysis without UV irradiation were analyzed, significant amounts of chlorinated contaminants were found in the samples. However, no contaminants were found in the end product of the UV/chlorinolysis experiment. Although no cost information was cited, the process successfully treated various hydrazines. The adaptability of the process to the treatment of other hazardous organics is currently unknown, although it would seem probable many other types of organics could be treated by the process. Careful monitoring for undesirable chlorinated by-products would be warranted when testing UV/Chlorinolysis with other organic compounds (59).

CATALYTIC HYDROGENATION-DECHLORINATION OF POLYCHLORINATED BIPHENYLS

Workers at the Osaka Prefectural Research Institute and the Daido Oxygen Company, Ltd., Japan, investigated several research parameters regarding the catalytic hydrogenation-dechlorination of PCB's. The work was reported in 1979. The authors investigated the use of both a Raney-Nickel catalyst and a carbon-suported palladium catalyst. Reaction of the PCB took place in an autoclave where PCB was dispersed and emulsified in an

aqueous sodium hydroxide solution containing isopropyl alcohol. Reaction
temperatures were kept constant within $\pm$ 0.5°C (60).

Experimental Details

The Raney-Nickel catalyst was prepared by reacting 10 grams of a 50-50
Nickel/Aluminum alloy powder in 30% aqueous sodium hydroxide solution for
50 minutes at 60°C. Great care was taken to ensure that the Raney-Nickel
catalyst's activity was constant. Preliminary bench-scale studies indicated the catalyst could be used more than five times. Experiments with
the Raney-Nickel catalyst were carried out in the autoclave with PCB's
such as Arochlor 1242 at temperatures between 70°C-200°C. A constant
volume solution of 100 mls composed of PCB, water, caustic, Raney-Nickel
catalyst, and isopropyl alcohol was reacted with hydrogen at either
100 kg/cm^2 or 30 kg/cm^2. By-products were measured by flame ionization
chromatography, electron capture chromatography for chlorinated molecules,
and a GC-mass spectrometer. Inorganic chloride was measured by silver
chloride gravimetry. Major by-products were biphenyls, phenylcyclohexane,
low levels of bicyclohexane, and sodium chloride.

Following the Raney-Nickel catalyst results, a series of experiments
using a carbon-supported palladium catalyst (5% palladium/3.3 wt% to
PCB KC-400) were conducted. PCB KC-400 is usually tetrachlorinated. Experiments were again carried out in an autoclave in a sodium hydroxide
aqueous solution containing isopropyl alcohol to disperse and emulsify
the PCB. Reactions were studied at constant near normal hydrogen pressures
of 25 kg/cm^2 at 50°-100°C.

Conclusions

The following preliminary conclusions were reached based on experimental evaluations:
- The rate of dechlorination of trichlorinated PCB Arochlor 1242
  was constant with constant hydrogen pressure until 65% dechlorination was achieved.
- An optimum alkaline ratio of 1, corresponding to 1.05 N NaOH was
  determined. An alkaline ratio of more than 1 (excess) inhibited
  the reaction, whereas an alkaline ratio of less than 1 completely
  stopped the reaction.
- The optimum concentration of the emulsifying dispersing agent
  isopropyl alcohol was found to be 30% by volume.
- PCB KC-400 was found to be 97% dechlorinated after 30 minutes
  reaction at 135°C.

Summary and Discussion

Assuming practically 100% dechlorination of KC-400 PCB, using a
Palladium-Carbon catalyst, the constant pressure hydrogenation/dechlorination
laboratory method for PCB disposal is promising, the authors noted. They
didn't cite any economic data or cost projections. Scale-up of the system
would be capital-intensive due to the cost of building a stainless steel
reactor, and the use of caustic would corrode the stainless steel reactor.

7

Economics of Emerging
Hazardous Waste Treatment Technologies

MOLTEN SALT COMBUSTION

No information was found in the literature for capital costs, nor operating plus maintenance costs regarding molten salt combustion as applied to hazardous wastes. Estimates from a representative of Anti-Pollution Systems indicate that it will cost from $50-$75 to burn 907 kg (1 ton) of municipal sludge (Personal communication from J. Greenberg to Dr. Barbara Edwards, Ebon Research Systems, July, 1979). Rockwell International scientists indicated that the cost to build a plant that burns hazardous waste at 45.4 kg/hr is $400,000-$500,000. To build a plant that burns 454 kg/hr would cost in the vicinity of $1.9-2.5 million (1979) (Personal communication from F. Rauscher to Dr. Barbara Edwards, Ebon Research Systems, June, 1979).

No firm cost estimates can be made until demonstration sized plants have functioned for an extended period of time. The economics of batch versus continuous type operations are also unknown. The type of hazardous waste combusted, the kind of melt utilized, and the material composition of the reactor vessel all will greatly influence cost considerations. Because the hot molten salt is corrosive, expensive stainless steel must be used as the reactor material. However, research is on-going to develop corrosion-resistant materials such as alumina and ceramics as reactor materials. This may reduce the cost of the reactor (Personal communication from F. Rauscher to John Paullin, Ebon Research Systems, September, 1980).

The capital costs of erecting a molten salt combustor will, of course, be quite high. Operating costs from the standpoint of fuel consumption are quite attractive when compared to conventional incineration. The use of auxiliary fuel to initially melt the bath may be necessary at start-up, depending on the type of waste combusted and its moisture content, yet the molten bath at operating temperatures can function as a heat sink, thereby requiring auxiliary fuel only at the onset. Heat from the combustion of wastes can be recycled. Because there are few moving parts in the system, maintenance costs would only be a small percentage of capital costs.

The treatment of bulk containers versus shredding the material requires further investigation. The use of bulk quantities for loading would probably result in lower costs.

Transportation costs for the removal of large amounts of spent melt (one ton or more) are unknown. Disposal in a conventional landfill may result in a leachate problem with the salts from the spent melt.

If particulate emissions are a problem, additional expense will be required for equipment to control the problem. Moreover, analytical chemistry instrumentation will be required to monitor the ambient air and the melt for residue levels of the hazardous waste or its possible harmful breakdown products.

To summarize, because molten salt technology as applied to the destruction of hazardous wastes is in its infancy, an economic cost profile is difficult to develop. The process is very efficient for combustion of certain wastes. The future will determine the economic viability of molten salt processing of hazardous wastes.

FLUIDIZED BED INCINERATION

In a paper presented at the American Chemical Society Meeting in September, 1979, Richard D. Ross of the Read-Ferry Company, Inc. stated that a fluidized bed incinerator is usually high in initial cost, yet when waste quantities are high, and the material cannot be handled in more conventional systems, fluidized bed incineration is practical (49). When contacted for more specific information on the cost of the process, Mr. Ross said that the cost has a wide range and depends on the type of waste, its caloric value, and the size of the combustor. A cost of 2-3 times more than a conventional incinerator equipped with a waste atomizing system would not be unusual. Problems with particulates requiring the addition of scrubbing systems would also increase costs. Costs for analytical instrumentation to monitor residue levels are also necessary (Personal communication from R.D. Ross to Dr. Barbara Edwards, Ebon Research Systems, Febuary, 1980).

According to M. Sittig, the type and composition of the waste is a significant design parameter that will impact cost not only during combustion but also during storage, processing, and transport prior to incineration (52). If the waste is a heterogeneous mixture, operations will be more complex, and the combustor will require auxiliary fuel. Homogeneous wastes that are injected and uniformly dispersed in the bed simplify the system design and cost less to incinerate. Installation and operating costs will vary significantly depending on the type of waste processed. Investment and operating costs have been estimated at approximately $20 and $5 per 907 kg (ton) respectively (1979). Maintenance costs, with no moving mechanical parts in the reactor would only be a small percentage of initial capital costs. Environmental control costs would be related to the type of equipment necessary to control particulate emissions and off-gases, e.g. cyclones and afterburners. Additional costs would be incurred for monitoring residues with analytical instrumentation (50). In some cases, depending on the character of the hazardous waste and the type of equipment available, conventional incinerators may be converted into fluidized bed combustors. This could represent a smaller capital investment compared to the cost of erecting a fluidized bed incinerator from the beginning (42).

Mr. Raymond Esposito operates a 45-cm tall, 0.35-kg/hour throughput fluidized bed facility for Union Chemical in Maine. The combustor can burn mixtures of municipal waste and chlorinated hydrocarbons with two seconds residence time using a sand bed. Mr. Esposito states that he hopes to scale up the unit so that it will process larger quantities at a cost of approximately $40 a drum. If approximately half the waste is municipal with a relatively high caloric content, it is not necessary to add oil as an auxiliary fuel (Personal communication from R. Esposito to Dr. Barbara Edwards, Ebon Research Systems, September, 1979).

Alternate fuel studies by the U.S. Army

During the evaluation of a pilot plant fluidized bed incinerator for the combustion of munitions, the U.S. Army studied the use of alternate fuels. Because large fluidized bed incinerators would have rather high pre-heat feul costs for liquid fuels low in sulfur and nitrogen, plus gaseous fuels are also expensive, studies were made regarding high sulfur and high nitrogen fuels by the U.S. Army Armament Research and Development Command (ARRADCOM) in Dover, New Jersey (41).

A 30 wt% nickel catalyst was evaluated for use with various fuels. Major results indicted satisfactory operation of a catalytic fluidized bed with both gaseous and low sulfur fuels. Initial tests revealed high sulfur fuels poisoned the nickel catalyst with subsequent high NO emissions, although stable combustion was obtained. Further investigation of various paramenters is necessary to fully evaluate the relationship of the nickel catalyst with high sulfur fuels.

Economic Projections by the U.S. Army

In addition to cost-savings studies for the use of alternate fuels, mathematical models were developed to compare the fluidized bed combustion of a 25 wt% TNT slurry versus the use of a rotary kiln. Projected costs savings of $19,000-$193,000 (1977) per year were estimated for a 112.5 kg/hr capacity fluidized bed system versus the rotary kiln. Cost savings of $108,000-$311,000 per year were projected for a larger 450 kg/hour capacity fluidized bed system. The primary factor in the cost-savings is lower operating cost (41).

Cost estimates were also made for a designed modular 202.5 kg (500 lb) per hour capacity fluidized bed system. A survey was made of major fluidized bed suppliers for the main components chamber, blower, cyclone, grid, instrumentation, and controls. The costs ranged from $210,000-$505,000 (1977). Based on these figures, a reasonable cost figure including design, fabrication, preliminary check-out, on-site supervision during assembly, operator training, and start-up assistance would be $400,000 (1977), according to the researchers. An additional $75,000 could be considered for an optional heat exchanger to preheat the fluidizing air.

Cost data for a TNT slurry preparation building containing two 1.89 m (500-gal) mixing tanks with pumps and pneumatic mixers, instrumentation, piping, design an construction, grinder, conveyors, building and miscellaneous items, the main fluidized bed system, and the optional heat exchanger were estimated at $685,000. The cost data are tabulated below:

Projected Costs 112.5 kg/hr TNT Slurry[#]
Fluidized Bed System (1877)

Main fluidized bed system	$400,000
Slurry preparation building	210,000
Heat exchanger	75,000
	$685,000

#) Source: R. Scola et al. U.S. Army ARRADCOM, Dover, New Jersey

Investment and Consumption Data for HCl Pickling Liquor

The following investment and consumption data were estimated by American Lurgi Corporation in 1972 based on a possible U.S. installation handling 30 metric tons/hr of spent HCl pickling liquor with a feed rate of 53 l/minute:

Investment $480,000 (1972)
Material Input and Output

Feed (1/min)	52.6
Rinse water (1/min)	50.3
Process water (1/min)	3.8
Power (watts)	165.0
Fuel (sm ^{3}/min)	4.9
Product HCl (18%–1/min)	54.6
Product iron oxide (kg/min)	7.0

This cost reflects engineering costs and equipment costs within the battery limits. Site preparation, building structure, and erection costs were not included (50).

Investment and Process Characteristics of a Blasting Abrasive Fluidized Bed

Personnel at the David Taylor Naval Ship Research and Development Center in Annapolis, Maryland, evaluated a fluidized bed for the combustion of spent blasting abrasive contaminated with organotin paint, in which the blasting abrasive served as the bed material. A design basis for a process model was established at a maximum feed rate of 45 metric tons/8 hr of free-drained abrasive containing 11 wt% water, and paint contaminant at 0.6%. A fluidizing air velocity of 75 m/minute was suggested. Paint particle concentration ratios of 5:1 and 10:1 were considered. At the 10:1 paint ratio, the process is autogenic; no auxiliary fuel is needed. Proposed design and cost data are cited in Table 15. Much lower costs are evident for the 10:1 paint ratio (51).

Consideration should also be given to laboratory costs incurred prior to scale—up. The efficiency of fluidized bed combustion technology for hazardous wastes depends on establishing ideal combustion parameters and a suitable med medium for the waste of interest.

To summarize, the economics for fluidized bed technology has been established for some hazardous wastes, and proposed for others. A fluid-ized bed can function as a heat sink, and heat recovery is possible. Based on these considerations, the application of fluidized bed combustion to certain types of hazardous waste can be considered economically viable.

TABLE 15

PROCESS CHARACTERISTICS FOR A PROPOSED BLASTING ABRASIVE FLUIDIZED BED (538°C)

Hours of Operation	Equipment Size Diameter, Meters Inside	Outside	Fuel/Power[*] Operating Costs $K (1977)	Estimated Captial Costs $K (1977)
		BULK ABRASIVE 1:1		
8	2.4	3.2	0.165	390
16	1.7	2.6	0.165	240
		CONCENTRATION 5:1		
8	10.8	2.0	0.024	175
16	5.1	5.1	0.024	125
		CONCENTRATION 10:1		
8	0.75	1.8	0.006[#]	180
16	0.24	1.3	0.006	115

(*) Assumes fuel costs of $15 per barrel, and power at 5¢ per kilowatt hour
(#) Assumes cost of concentrating operation at $45,000 (1977)

Source: A. Ticker et al. David Taylor Naval Ship Research
              and Development Center.

UV/OZONE DESTRUCTION

Many complex decisions representing trade—offs are necessary to im-plement a well—designed, efficiently operated, economic UV/ozonation system. UV/ozonation is generally restricted to waste streams of 1% or lower hazardous contaminant. Because ozone is non—selective as an oxidant, the waste stream should primarily contain the waste of interest. Effluent streams must be carefully monitored for toxic intermediates. This will add to the cost of the process. Modern systems are usually automated, thereby requiring lower labor costs.

Ozone Generation

Ozone must be generated on site to treat hazardous waste, because there is no practical method to store or ship it. Most ozone generators employ high voltage, although equipment for small outputs is available. Ozone generation using oxygen is about twice as efficient as air, yet oxygen costs about twice as much as air. Ozone generators can be air cooled or water cooled, yet experience has shown than the more expensive water cooling system has less problems and is a more trouble free system according to Klein et al. (25).

Power

The major operating cost for ozone manufacturing is the cost of electric power. Of the total power utilized in ozone generation, approximately 10% actually produces the ozone. The rest is consumed in air handling, air preparation, and waste heat. Typical power consumption figures in plants using air as the generator range from 6-8 kwh/lb ozone for the ozone generator alone, and 10-13 kwh/lb ozone total consumption including air handling and preparation. With oxygen as the feed gas for generating ozone, typical power consumption figures range from 3-4 kwh/lb ozone for ozone generation, and 7-12 kwh/lb ozone total consumption (30). Although higher frequency generating systems are more efficient ozone generators, the electrical cost of the high frequency can be considered to offset the gain. New advances in solid state technology may improve the situation (5). Ozonators which operate at high pressures require extra electrical power to operate high pressure compressors. According to the Ozone Research and Equipment Corporation, the cost of additional electric power is lower than the labor cost and inconvenience associated with periodic tear down, cleaning, and reassembly of the generators (27).

Destruct Systems

In some cases, ozone not consumed in the primary reactor system can be reused or applied in another system. It may be necessary to provide a destruct system to guarantee removal of any unused ozone from the system before the gas stream is discharged to the atmosphere. There is a choice of three types of destruct systems thermal, catalytic, and combination. Thermal systems heat the entire gas stream to a high temperature for a specific time frame. Catalytic systems heat the gas to 250°C and then pass the gas stream through a solid phase catalyst bed for ozone destruction. The catalysts are proprietary. Combination systems exist. These systems represent a trade-off between heating costs and catalysts costs. Heat recovery should receive consideration (30).

UV Lamps

The cost for a bank of UV lamps is small compared to the other capital costs of an UV/ozonation system. Increased UV light can favorably accelerate ozone decomposition, but at a trade-off in higher cost requirements. Over a period of time, a thin film builds up on the UV lamps and impairs their efficiency. Consideration must be given to implementing a system to wipe the lamps or use extra labor. Because of the wide variation in reactivities of organic contaminants, UV levels cannot be generalized, and should be determined for each waste of interest. Since ozone is more stable in acidic solutions, neutralization may seriously affect ozone levels, thereby affecting UV requirements, and reducing cost-effectiveness of the system.

Capital Costs and Operating Costs

Predicted capital and operating costs for a proposed pilot plant treating 150,000 gallons per day of PCB's at levels of 50 ppm reduced to values less than or equal to 1 ppm are shown in Table 16 (48). Pertinent data comparing a 40,000 and 150,000 gallons per day plant treating PCB's at feed levels of 50 ppm reduced to 1 ppm PCB are shown in Table 17 (48).

Summary of UV/Ozonation Economics

The data presented here demonstrate the economic viability for an UV/ozone system to treat certain hazardous wastes. Typically capital costs for scale-up are not directly proportional to the increased treatment capacity, e.g., a system scaled up to 3.75 times as large costs only 2.4 times more. The daily operating costs for the larger (3.75X) system versus the smaller system are also not proportional. The 3.75X system only costs nearly 2X the smaller system on a daily operating basis.

TABLE 16
MINIMUM OPERATING AND CAPITAL COSTS FOR A 150,000 GPD ULTROX TREATMENT[*]
PILOT PLANT TO OBTAIN $\leq$ 1 ppm PCB

150,000 Gallons per Day Plant (567,750 liters/day)

Total Capital Costs $	Operating Cost $/1,000 gal	PCB levels ppm	Ozone wt%	Total Number of Lamps
350,280	1.91	0.7	1.0	15
300,320	1.71	0.9	1.0	15
300,320	1.71	0.9	1.0	15
300,320	1.71	0.9	1.0	15
300,320	1.71	1.0	1.0	15

(*) Source: Arisman and Musick, General Electric Company, Zeff and Crase, Westgate Research Corporation, West Los Angeles, California (1980).

TABLE 17
DESIGN SPECIFICATIONS, CAPITAL, AND O&M COSTS FOR [*]
40,000 AND 150,000 GPD ULTROX TREATMENT PLANTS
(50 ppm PCB feed-1 ppm PCB effluent)

DESIGN SPECIFICATIONS

Reactor	40,000 GPD (151,400 LPD) Automated System	150,000 GPD (567,750 LPD) Automated System
Dimension, Meters (LxWxH)	2.5 x 4.9 x 1.5	4.3 x 8.6 x 1.5
Wet Volume, Liters	14,951	56,018
UV Lamps; Number 65 W	378	1179
Total Power, kw	25	80

Ozone Generator

	40,000 GPD	150,000 GPD
Dimensions, Meters (LxWxD)	1.7 x 1.8 x 1.2	2.5 x 1.8 x 3.1
kg Ozone/day	7.7	28.6
Total Energy required (kwh/day)	768	2544

BUDGETARY EQUIPMENT PRICES

	40,000 GPD	150,000 GPD
Reactor	$94,500	$225,000
Generator	30,000	75,000
TOTAL	$124,500	$300,000

O & M Costs/Day

	40,000 GPD	150,000 GPD
Ozone Generator Power	$4.25	$15.60
UV Lamp Power	15.00	48.00
Maintenance (Lamp Replacement)	27.00	84.20
Equipment Amortization (10 Yrs @ 10%)	41.90	97.90
Monitoring Labor	85.71	85.71
TOTAL/DAY	$173.86	$331.41
Cost per 3785 Liters (with monitoring labor)	$4.35	$2.21
(without monitoring labor)	$2.20	$1.64

(*) Source: Arisman and Musick, General Electric Company, Zeff and Crase
Westgate Research Corporation, West Los Angeles, California (1980)

8

Survey of Hazardous Waste Generators

PURPOSE

In conjunction with the assessment of emerging hazardous waste destruction technologies, Ebon Research Systems conducted a survey of both large and small hazardous waste generators. The aims of the survey were:
- To determine user needs for the application of treatment techniques to pollutant disposal problems.
- To determine if the hazardous waste generators had access to, or desired access to computerized services and/or a newsletter regarding hazardous waste disposal.
- To inform hazardous waste generators that new emerging techniques are available for hazardous waste destruction.

Ebon Research Systems was also interested in compiling the response, and determining the overall level of interest from those companies surveyed.

SURVEY RESULTS

The instrument for the survey is show in Figure 35. Except for question 6 which asked for additional information, the replies were limited to a yes or no. Respondents were informed that the names of individual companies would be kept confidential. Ebon Research Systems believed this simple approach would be more likely to elicit a higher response level.

The survey was mailed in January, 1980, to the 53 chemical companies participating in the waste disposal site survey conducted by the House of Representatives Subcommittee on Oversight and Investigation, of the Committee on Interstate and Foreign Commerce, 96th Congress, First Session. The Committee derived the list of participants (considered to be the 53 largest companies in terms of domestic sales) from the 1976 Kline Guide to the Chemical Industry, and the American Chemical Society's 1977 listing of the top 50 chemical companies. Because the initial response to the survey was low, more companies were added to the list by Ebon Research Systems. Even though the mailing was followed up by telephone request, the total number of companies responding was 31 out of 73 contacted. Of these companies, only 18 were on the original Subcommitte's listing of 53.

The response matrix relating to user needs is shown in Table 18.

123

TABLE 18
MATRIX RELATING TREATMENT TECHNIQUES TO USER NEEDS

Treatment Technique	% of Users Requesting Information
Conventional incineration	68
Fluidized bed combustion	55
Secured landfills	84
Ocean incineration	32
Ozonation	32
Chlorinolysis	26
Molten salt combustion	32
UV destruction	35

The following observations were pertinent:

- Only one company was interested in landfarming.
- The most interest was expressed in conventional established technologies.
- Sixty-five percent of respondents already subscribed to a newspaper discussing hazardous waste at least in part, but only 19% had access to a computerized service regarding hazardous waste disposal, at least in part.
- Fifty-eight percent indicated they would use both a computerized data bank and newsletter regarding hazardous waste, while 16% stated they would not be interested in either.
- More than 80% of the companies expressed interest in more information regarding hazardous waste legislation, economics of waste disposal and transport, spill cleanup techniques, and the technology for disposal or reclamation.

Eighteen companies responded to Question 6: "Is there any other way that a hazardous waste information service could be of use to your company?" The major interests are summarized below:

- Names and locations of authorized hauling contractors
- Permit information for companies which operate in more than one state
- Waste exchange and recycling information
- Packaging and transportation regulation information
- Strongly publicize significant hazardous waste control measures implemented by industry
- Computerized data bases are needed for information on aerobic or anaerobic degradation of organic chemicals.
- Btu or fuel values of combustible products are needed so that an economic assessment of various disposal practices can be made.

FIGURE 35

EBON RESEARCH SYSTEMS' HAZARDOUS WASTE INFORMATION SERVICE QUESTIONNAIRE

Please return to:
Barbara H. Edwards,
1542 9th St., N.W.
Washington, D.C. 2001

Please indicate responses with an X.	YES	NO

1. Does your company now have access to
    computerized services (data banks, etc.)
    on hazardous waste disposal?

2. Does your company currently subscribe
    to a news letter on hazardous waste disposal?

3. If the USEPA established a data bank or
    quarterly newsletter on hazardous waste
    disposal, which would be of most use to your
    your company?

    (a) data bank only

    (b) newsletter only

    (c) both data bank and newsletter

    (d) would not use either data bank or newsletter

4. Please indicate any of the following topics
    that would be of interest to your company

    (e) hazardous waste legislation

    (f) economics of hazardous waste disposal
         and transport

    (g) hazardous waste spill cleanup techniques

    (h) technology of hazardous waste disposal
         reclamation

5. Would up-to-date information of any of the
    specific technologies listed below be of
    interest to your company?

    (i) conventional incineration

    (j) fluidized bed combustion

FIGURE 35 (Continued)

EBON RESEARCH SYSTEMS' HAZARDOUS WASTE INFORMATION SERVICE QUESTIONNAIRE

Please indicate responses with an X.	YES	NO
(information on specific technologies-cont.)		
(k) secured landfills		
(l) ocean incineration		
(m) ozonation		
(n) chlorinolysis		
(o) molten salt combustion		
(p) UV destruction		
(q) other (please specify)		

6. Is there any other way that a hazardous waste information waste information service could be of use to your company?

_____

_____

_____

Use the back of the questionnaire or another sheet for further comments.

THANK YOU

References

1. Greenberg, J. 1972. Method of Catalytically Inducing Oxidation of Carbonaceous Materials by the Use of Molten Salts. United States Patent 3,647,358, assigned to Anti-Pollution Systems, Inc., Pleasantville, New Jersey, 16 pp.

2. Greenberg, J. and D.C. Whitaker. 1972. Treatment of Sewage and Other Contaminated Liquids With Recovery of Water by Distillation and Oxidation. United States Patent 3,642,583, assigned to Anti-Pollution Systems, Inc., Pleasantville, New Jersey, 10 pp.

3. Findlay, A. et al. 1951. Phase Rule. Dover Publications, Inc., New York, New York, 494 pp.

4. Greenberg, J. 1979. The use of molten salts in emission control. Presented at the 72nd Annual Meeting of the Air-Pollution Control Association, Cincinnati, Ohio, June 24-29.

5. Heredy, L.A. et al. 1969. Removal of Sulfur Oxides From Flue Gas. United States Patent 3,438,722, assigned to North American Rockwell Corporation, California, 12 pp.

6. Yosim, S.J. et al. 1979. Molten salt destruction of hazardous wastes produced in the laboratory. In: Safe Handling of Chemical Carcinogens, Mutagens and Teratogens: The Chemists Viewpoint, Symposium held at the American Chemical Society, Hawaii.

7. Yosim, S.J. et al. 1979. Disposal of hazardous wastes by molten salt combustion. In: Ultimate Disposal of Hazardous Waste, Symposium held at the American Chemical Society, Hawaii.

8. Yosim, S.J. et al. 1978. Destruction of pesticide and pesticide containers by molten salt combustion. In: American Chemical Society Symposium Series, No. 73, Disposal and Decontamination of Pesticides, M.V. Kennedy, ed., pp. 118-130

9. Smithson, G.R. 1977. Utilization of energy from organic wastes through fluidized bed combustion. In: Fuels from Waste, L.L. Anderson and D.A. Fullman, eds., Academic Press, New York, New York. pp. 195-209

10. Botterill, J.S. 1977. The contribution of fluid-bed technology to energy saving and environmental protection. Applied Energy, 3:139-150.

11. Yerushalmi, J. and N.T. Cankurt. 1978. High-velocity fluid beds. Chemtech., 8(9):564-571.

12. Oehlschlaeger, H.F. 1976. Reactions of ozone with organic compounds. Proceedings of an Ozone Conference, Cincinnati, Ohio.

13. Bailey, Philip S. 1973. Activity of ozone with various organic functional groups important to water purification. Presented before the First International Symposium on Ozone for Water and Wastewater Treatment.

14. Prengle, H.W. and C.E. Mauk. 1978. New technology: ozone/UV chemical oxidation wastewater process for metal complexes, organic species and disinfection. The American Institute of Chemical Engineers Symposium Series, 74:228-243.

15. Prengle, H.W. et al. 1975. Ozone/UV process effective wastewater treatment. Environmental Management, 54:82-87.

16. Lee, M.K. 1980. Study of UV-ozone reactions with organic compounds in water. Presented Before the Division of Environmental Chemistry, American Chemical Society, Houston, Texas.

17. Yosim, S.J. and K.M. Barclay. 1978. Destruction of hazardous wastes by molten salt combustion. In: Proceedings of a National Conference About Hazardous Waste Management, San Francisco, California, pp. 146-156.

18. Wilkinson, R.R. et al. 1978. State-of-the-Art-Report, Pesticide Disposal Research. EPA-60012-78-183, U.S. Environmental Protection Agency, Cincinnati, Ohio, 225 pp.

19. Kahl, A.L. et al. 1978. The molten salt coal gasification process. Chemical Engineering Progress, August, pp. 73-79.

20. Yosim, S.J. et al. 1973. Non-polluting disposal of explosives and propellants. United States Patent 3,778,320, assigned to Rockwell International Corporation, California.

21. Yosim, S.J. et al. 1974. Disposal of Organic Pesticides. United States Patent 3,845,190, assigned to Rockwell International Corporation, El Segundo, California, 10 pp.

22. Grantham, L.F. et al. 1979. Disposal of PCB and other toxic hazardous waste material by molten salt combustion. In: National Conference on Hazardous Risk Assessment, Disposal and Management Proceedings, Miami, Florida, pp. n.s.

23. Dustin, D.F. et al. 1977. Applications of molten salt incineration to the demilitarization and disposal of chemical material. EM-TR-76099, Department of the Army, Edgewood Arsenal, Aberdeen Proving Ground, Maryland, 55 pp.

24. Powers, P.W. 1976. How to Dispose of Toxic Substances and Industrial Wastes. Noyes Data Corporation, Park Ridge, New Jersey. 497 pp.

25. Klein, M.J. et al. 1973. Generation of ozone. Presented before the
     First International Symposium on Ozone for Water and Wastewater
     Treatment.

26. Anonymous. Comparative Data. Ozone Generators and Air Pre-Drying
     Equipment, brochure, Ozonair Corporation, San Francisco, California.

27. Anonymous. Ozone Technology, brochure 124, Ozone Research and Equipment
     Corporation, Phoenix, Arizona.

28. Bollyky, L.J. 1973. Ozone treatment of cyanide and plating wastes.
     Presented before the First International Symposium for Water and
     Wastewater Treatment.

29. Augugliaro, V. and L. Rizzuti. 1978. The pH dependence of the ozone
     absorption kinetics in aqueous phenol solutions. Chemical Engineering
     Science 33:1141-1147.

30. Derrick, D.H. and J.R. Perrich. 1979. Guide to ozone equipment
     selection. Pollution Engineering, 11:42-44.

31. Hammond, V.L. and L.K. Mudge. 1975. Feasibility Study of Use of Molten
     Salt Technology for Pyrolysis of Solid Waste. EPA-67012-75-014,
     Prepared by Battelle Pacific Northwest Laboratories for the U.S.
     Environmental Protection Agency, Cincinnati, Ohio, 81 pp.

32. Bliss, C. and B.M. Williams, eds. 1977. Proceedings of the Fifth
     International Conference on Fluidized Bed Combustion, Washington, D.C.

33. Wall, C.J. et al. 1975. How to burn salty sludges. Chemical
     Engineering, 80(8): 77-82.

34. Landreth, R.E., and C.J. Rogers. 1974. Fluidized bed combustion of
     phenol and methyl methacrylate wastes. In: Promising Technologies for
     Treatment of Hazardous Wastes. EPA 670/2-74-088, U.S. Environmental
     Protection Agency, Cincinnati, Ohio. pp. 78-87.

35. Ragland, K.W. and D.P. Paul. 1979. Fluidized bed combustion of plastic
     with coal. Presented at the Fourteenth Intersociety Energy Conversion
     Conference, Boston, Massachusetts.

36. Kamino, Y. et al. 1978. Gasification of waste plastics. Technical
     Report of the Hitachi Ship Building Technology Research Institute,
     Japan. pp. 16-21.

37. Eggers, F.W. et al. 1977. Removing Chlorine-Containing Insulation with
     a Fluidized Medium Containing Reactive Calcium Compounds. United
     States Patent 4,040,868, assigned to Cerro Corporation, New York,
     New York. 8 pp.

38. Walker, W.M. 1973. Fluid bed incineration of chlorinated hydrocarbons.
     Australian Mines Development Laboratory Bulletin, 16:41-44.

39. Ziegler et al. 1973. Fluid bed incineration. RFP-2016, Technical Report of Dow Chemical U.S.A., Rocky Flats Division, Golden, Colorado. 13 pp.

40. Ziegler et al. 1974. Pilot plant development of a fluidized bed incineration process. RFP-227, Technical Report of Dow Chemical U.S.A., Rocky Flats Division, Golden, Colorado. 10 pp.

41. Scola, R. et al. 1978. Fluidized bed incinerator for disposal of propellants and explosives. Technical Report of the U.S. Armament Research and Development Command, Dover, New Jersey. 114 pp.

42. Carroll, J.W. et al. 1979. Assessment of hazardous air pollutants from disposal of munitions in a prototype fluidized bed incinerator. American Industrial Hygiene Association Journal, 40:147-158.

43. Wong, A.S. et al. 1979. Ozonation of 2,3,7,8-Tetrachlorodibenzo-p-Dioxin. Presented at the Symposium on the Chemistry of Chlorinated Dibenzodioxins and Dibenzofurans, ACS National Meeting, Washington, D.C.

44. Sierka, R.A. and W.F. Cowen. 1980. The catalytic ozone oxidation of aqueous solutions of hydrazine, monomethyl hydrazine and unsymmetrical dimethylhydrazine. Presented at the 35th Annual Purdue Industrial Waste Treatment Conference, Layfayette, Indiana.

45. Cowen, W.F. et al. 1979. Identification of the partial oxidation products of hydrazine, monomethyl hydrazine and unsymmetrical dimethyl hydrazine from ozonation. Paper presented at the USAF Engineering and Service Center, Tyndall Air Force Base, Florida.

46. Leitis, E. et al. 1979. An investigation into the chemistry of the UV/Ozone purification process. Presented at the 4th World Ozone Congress, Houston, Texas.

47. Macur, G.J. et al. 1980. Oxidation of organic compounds in concentrated industrial waste water with ozone and ultraviolet light. Presented at the 35th Annual Purdue Industrial Waste Conference, Layfayette, Indiana.

48. Arisman, R.K. and R.C. Musick. 1980. Experience in operation of a UV-Ozone (ULTROX) pilot plant for destroying PCB's in industrial waste effluent. Presented at the 35th Annual Purdue Industrial Waste Conference, May, 1980.

49. Ross, R.D. 1979. Incineration—a positive solution to hazardous waste disposal. Presented at The American Chemical Society Division of Chemical Health and Safety Meeting, Washington, D.C.

50. Marnell, P. 1972. Spent HCl pickling liquor regenerated in a fluid bed. Chemical Engineering, 79(25):102-103.

51. Ticker, A. et al. 1979. Study of the fluidized bed process for treatment of spent blasting abrasives. Journal Coating Technology, 49(626):29-35.

52. Sittig, M. 1979. Incineration of Industrial Hazardous Wastes. Noyes Data Corporation, Park Ridge, New Jersey. 348 pp.

53. Landreth, R.E. and C.J. Rogers. 1974. Wet air oxidation. In: Promising Technologies for Treatment of Hazardous Wastes. EPA-670/2-74-088, U.S. Environmental Protection Agency, Cincinnati, Ohio.

54. Dresen, R.W. and J.R. Mayer. 1972. Wet Combustion of Organics. United States Patent 3,984,311, assigned to Dow Chemical Company.

55. Miller, R.A. et al. 1980. Destruction of toxic chemicals by catalyzed wet oxidation. Presented at the Purdue Industrial Waste Conference, West Layfayette, Indiana.

56. Kitchens, J.A. 1979. Dehalogenation of Halogenated Compounds. United States Patent 4,144,152, assigned to Atlantic Research Corporation, Alexandria, Virginia.

57. Jones, W.E. et al. 1980. Light activated reduction of chemicals. Presented before the Division of Environmental Chemistry, American Chemical Society, Houston, Texas.

58. Trump, J.G. et al. 1979. Destruction of pathogenic microorganisms and toxic organic chemicals by electron treatment. Presented at the Eigth National Conference on Municipal Sludge Management, Miami, Florida.

59. Fochtman, E.G. et al. 1979. Chlorinolysis treatment of hydrazine in dilute aqueous solution. Presented at Symposium of Environmental Chemistry of Hydrazine, Tyndall Air Force Base, Florida.

60. Yasuhiro, H. et al. 1979. Conversion of polychlorinated biphenyls to useful materials by a catalytic hydrogenation-dechlorination method. Presented at the American Chemical Society Meeting in Honolulu, Hawaii.

Appendix A
Eutectic Mixtures of Neutral Salts

Mixtures with KCl	Eutectic Temperature
41 M% KCl-59 M% LiCl	358°C
57 M% KCl-43 M% BaCl$_2$	345°C
60 M% KCl-40 M% CaCl$_2$	580°C
35 M% KCl-65 M% CdCl$_2$	380°C
40 M% KCl-60 M% MgCl$_2$	420°C
55 M% KCl-45 M% SrCl$_2$	575°C
50 M% KCl-50 M% MnCl$_2$	500°C
48 M% KCl-52 M% PbCl$_2$	411°C
45 M% KCl-55 M% ZnCl$_2$	230°C

Mixtures with LiCl	Eutectic Temperature
72 M% LiCl-28 M% NaCl	560°C
28 M% LiCl-62 M% CaCl$_2$	496°C
45 M% LiCl-55 M% MnCl$_2$	550°C
45 M% LiCl-55 M% PbCl$_2$	410°C
45 M% LiCl-55 M% SrCl$_2$	475°C

Mixtures with Barium Chloride	Eutectic Temperature
12 M% BaCl$_2$-88 M% BeCl$_2$	390°C
30 M% BaCl$_2$-70 M% CaCl$_2$	600°C

APPENDIX A (Continued)

EUTECTIC MIXTURES OF NEUTRAL SALTS

Mixtures with Lead Dichloride	Eutectic Temperature
20 M% $BaCl_2$–80 M% $PbCl_2$	505°C
48 M% $BeCl_2$–52 M% $PbCl_2$	300°C
90 M% $BiCl_2$–10 M% $PbCl_2$	205°C
18 M% $CaCl_2$–82 M% $PbCl_2$	460°C
35 M% $CdCl_2$–65 M% $PbCl_2$	387°C
33 M% CuCl–67 M% $PbCl_2$	285°C
50 M% $FeCl_3$–50 M% $PbCl_2$	185°C
48 M% KCl–52 M% $PbCl_2$	411°C
45 M% LiCl–55 M% $PbCl_2$	410°C
8 M% $MgCl_2$–92 M% $PbCl_2$	450°C
30 M% $MnCl_2$–70 M% $PbCl_2$	405°C
28 M% NaCl–72 M% $PbCl_2$	415°C
72 M% PbF_2–28 M% $PbCl_2$	550°C
76 M% PbI_2–24 M% $PbCl_2$	310°C
50 M% $SnCl_2$–50 M% $PbCl_2$	410°C
60 M% TlCl–40 M% $PbCl_2$	390°C
50 M% $ZnCl_2$–50 M% $PbCl_2$	340°C

Mixtures with Magnesium Chloride	Eutectic Temperature
8 M% $MgCl_2$–92 M% $PbCl_2$	460°C
55 M% $MgCl_2$–45 M% $SrCl_2$	530°C

APPENDIX A (Continued)

EUTECTIC MIXTURES OF NEUTRAL SALTS

Mixtures with Sodium Chloride	Eutectic Temperature
33 M% NaCl-67 M% CaCl$_2$	550°C
42 M% NaCl-58 M% CdCl$_2$	397°C
48 M% NaCl-52 M% CoCl$_2$	365°C
60 M% NaCl-40 M% MgCl$_2$	450°C
55 M% NaCl-45 M% MnCl$_2$	415°C
45 M% NaCl-55 M% NiCl$_2$	560°C
28 M% NaCl-72 M% PbCl$_2$	415°C
50 M% NaCl-50 M% SrCl$_2$	560°C

Mixtures with Calcium Chloride	Eutectic Temperature
47 M% CaCl$_2$-63 M% MnCl$_2$	600°C
18 M% CaCl$_2$-82 M% PbCl$_2$	470°C
50 M% CaCl$_2$-50 M% ZnCl$_2$	600°C

Mixtures with Cadmium Chloride	Eutectic Temperature
50 M% CdCl$_2$-50 M% MnCl$_2$	600°C
35 M% CdCl$_2$-65 M% PbCl$_2$	387°C
60 M% CdCl$_2$-40 M% SrCl$_2$	500°C
50 M% CdCl$_2$-50 M% ZnCl$_2$	500°C

Mixtures with Zinc Chloride	Euctectic Temperature
38 M% ZnCl$_2$-62 M% SnCl$_2$	180°C
45 M% ZnCl$_2$-55 M% SrCl$_2$	480°C

APPENDIX A (Continued)

EUTECTIC MIXTURES OF NEUTRAL SALTS

Mixtures with Potassium Bromide	Eutectic Temperature
31 M% KBr-69 M% LiBr	310°C
60 M% KBr-40 M% $CdBr_2$	325°C
32 M% KBr-68 M% $MgBr_2$	350°C
50 M% KBr-50 M% $SrBr_2$	525°C

Mixtures with Sodium Bromide	Eutectic Temperature
40 M% $BaBr_2$-60 M% NaBr	600°C
60 M% $CaBr_2$-40 M% NaBr	510°C
45 M% $CdBr_2$-55 M% NaBr	370°C
40 M% $MgBr_2$-60 M% NaBr	425°C
60 M% $SrBr_2$-40 M% NaBr	480°C

Mixtures with Lead Dibromide	Eutectic Temperature
80 M% $BiBr_3$-20 M% $PbBr_2$	200°C
18 M% $CdBr_2$-82 M% $PbBr_2$	344°C
90 M% $HgBr_2$-10 M% $PbBr_2$	208°C
50 M% $PbCl_2$-50 M% $PbBr_2$	425°C
78 M% PbF_2-22 M% $PbBr_2$	520°C
10 M% PbF_2-90 M% $PbBr_2$	350°C
44 M% PbI_2-56 M% $PbBr_2$	282°C

Mixtures with Lithium Bromide	Eutectic Temperature
80 M% LiBr-20 M% NaBr	525°C
75 M% LiBr-25 M% $BaBr_2$	485°C

APPENDIX A (Continued)

EUTECTIC MIXTURES OF NEUTRAL SALTS

Mixtures with 3 Chlorides	Eutectic Temperature
35 M% $PbCl_2$–35 M% KCl–30 M% $CdCl_2$	328°C
60 M% KCl–20 M% $PbCl_2$–20 M% NaCl	500°C
10 M% $PbCl_2$–50 M% KCl–40M% $ZnCl_2$	280°C
10 M% NaCl–40 M% LiCl–50 M% KCl	400°C
70 M% LiCl–15 M% $CaCl_2$–15 M% KCl	450°C
30 M% $BaCl_2$–35 M% $CaCl_2$–55 M% KCl	542°C
10 M% NaCl–35 M% $CaCl_2$–55 M% KCl	600°C
15 M% NaCl–50 M% $CdCl_2$–35 M% KCl	450°C
10 M% NaCl–15 M% $PbCl_2$–75 M% KCl	500°C

Mixtures with 3 Salts	Eutectic Temperature
12.5 M% NaBr–12.5 M% $CdBr_2$–75 M% $PbBr_2$	280°C
65 M% PbI_2–18 M% $PbCl_2$–17 M% $PbBr_3$	300°C

Mixtures with 4 Salts	Eutectic Temperature
25 M% $CdCl_2$–25 M% $CdBr_2$–25 M% $PbCl_2$–25 M% $PbBr_2$	450°C
10 M% NaCl–10 M% NaBr–60 M% $PbCl_2$–20 M% $PbBr_2$	450°C
10 M% NaCl–10 M% NaI–40 M% $PbCl_2$–40 M% PbI_2	365°C

Appendix B
Eutectic Mixtures of Active Salts

Oxides	Lowest Melt Temperature
43 M% WO_3–57 M% K_2WO_4	600°C
47 M% MoO_3–53 M% Li_2MoO_4	530°C
45 M% 3 $Na_2O \cdot As_2O_5$–55 M% As_2O_5	570°C
77 M% MoO_3–23 M% Na_2MoO_4	510°C
38 M% WO_3–62 M% Rb_2WO_4	570°C
8 M% CaO–92 M% P_2O_5	409°C
50 M% PbO–50 M% V_2O_5	480°C

Three Oxides	Lowest Melt Temperature
37 M% $K_2MoO_4(K_2O + MoO_3)$–63 M% Li_2MoO_4	525°C
50 M% KPO_3–50 M% $LiPO_3$ (Consists of K_2O, Li_2O, P_2O_5)	562°C
49 M% $NaPO_3$–49 M% KPO_3–2 M% K_2O	547°C
50 M% Li_2MO_4–50 M% Na_2MoO_4	465°C
50 M% Na_2WO_4–50 M% Li_2WO_4	500°C
50 M% Pb_2SiO_4–50 M% Na_2SiO_3	600°C
33 M% $Na_2O \cdot SiO_2$–67 M% $PbO : SiO_2$	580°C

Perchlorates	Lowest Melt Temperature
70 M% $LiClO_4$–30 M% NH_4ClO_4	185°C
40 M% $NaClO_4$–60 M% $Ba(ClO_4)_2$	305°C

APPENDIX B (Continued)

EUTECTIC MIXTURES OF ACTIVE SALTS

Sulfate Mixtures	Lowest Melt Temperature
33 M% K_2SO_4-67 M% $LiPO_3$	470°C
25 M% K_2SO_4-25 M% Li_2SO_4-25 M% KNO_3-25 M% $LiNO_3$	460°C
25 M% K_2SO_4-25 M% Li_2SO_4-25 M% K_2WO_4 25 M% Li_2WO_4	570°C
25 M% Li_2SO_4-25 M% Na_2SO_4-25 M% Li_2MoO_4-25 M% Na_2MoO_4	520°C

Nitrates	Lowest Melt Temperature
50 M% KNO_3-50 M% $NaNO_3$	222°C
35 M% KNO_3-65 M% $TlNO_3$	185°C
85 M% KNO_3-15 M% $Ba(NO_3)_2$	280°C
50 M% $Ca(NO_3)_2$-50 M% $NaNO_3$	240°C
50 M% $LiNO_3$-50 M% $Ba(NO_3)_2$	400°C
50 M% $Ca(NO_3)_2$-50 M% KNO_3	240°C
70 M% $LiNO_3$-30 M% $Ca(NO_3)_2$	240°C
50 M% $Ba(NO_3)_2$-50 M% $Ca(NO_3)_2$	520°C
60 M% $Tl(NO_3)_2$-40 M% $Ca(NO_3)_2$	140°C
40 M% KNO_3-35 M% $LiNO_3$-25 M% $NaNO_3$	130°C
60 M% KNO_3-25 M% $LiNO_3$-15 M% $Ca(NO_3)_2$	125°C
20 M% $LiNO_3$-60 M% $Cd(NO_3)_2$-20 M% KNO_3	140°C
34 M% $Ba(NO_3)_2$-33 M% KNO_3-33 M% $NaNO_3$	450°C
34 M% $Ca(NO_3)_2$-33 M% KNO_3-33 M% $NaNO_3$	140°C
34 M% $Cd(NO_3)_2$-33 M% $NaNO_3$-33 M% $LiNO_3$	176°C
5 M% Na_2SO_4-95 M% $NaNO_3$	300°C

APPENDIX B (Continued)

EUTECTIC MIXTURES OF ACTIVE SALTS

Hydroxides	Lowest Melt Temperature
50 M% $NaNO_3$–50 M% NaOH	270°C
50 M% KNO_2–50 M% KOH	200°C
50 M% KOH–50 M% KNO_3	235°C
70 M% KOH–30 M% LiOH	110°C
50 M% KOH–50 M% $NaNO_3$	240°C
50 M% KOH–50 M% NaOH	170°C
15 M% K_2CrO_4–85 M% KOH	361°C
70 M% NaOH–30 M% LiOH	210°C
50 M% NaOH–50 M% $NaNO_2$	266°C
50 M% KNO_3–50 M% NaOH	330°C
62 M% $Ba(OH)_2$–38 M% $Sr(OH)_2$	360°C
30 M% KOH–30 M% NaOH–40 M% LiOH	300°C
33 M% NaOH–34 M% Na_2SO_4–33 M% NaCl	500°C
25 M% LiOH–25 M% Li_2CrO_4–25 M% NaOH–25 M% Na_2CrO_4	475°C
25 M% LiOH–25 M% NaOH–25 M% $LiNO_3$–25 M% $NaNO_3$	400°C

Sulfates	Lowest Melt Temperature
60 M% K_2SO_4–40 M% $CoSO_4$	540°C
55 M% K_2SO_4–45 M% $CuSO_4$	460°C
55 M% Li_2SO_4–45 M% $CdSO_4$	575°C
60 M% Li_2SO_4–40 M% $CoSO_4$	600°C
35 M% $MnSO_4$–65 M% Li_2SO_4	600°C
50 M% $ZnSO_4$–50 M% Na_2SO_4	500°C

APPENDIX B (Continued)

EUTECTIC MIXTURES OF ACTIVE SALTS

Nitrites	Lowest Melt Temperature
70 M% $NaNO_2$–30 M% KNO_2	230°C

Nitrites-Nitrates	Lowest Melt Temperature
50 M% KNO_2–50 M% KNO_3	350°C
62 M% KNO_2–38 M% $Ca(NO_3)_2$	140°C
50 M% $NaNO_2$–50 M% KNO_3	150°C
50 M% $NaNO_2$–50 M% $NaNO_3$	230°C
50 M% $NaNO_2$–50 M% $Ca(NO_3)_2$	190°C
25 M% $NaNO_2$–25 M% KNO_2–25 M% $NaNO_3$–25 M% KNO_3	150°C
25 M% KNO_2–25 M% KNO_3–25 M% $Ba(NO_3)_2$–25 M% $Ca(NO_3)_2$	300°C

Appendix C
Hazardous Wastes Destroyed
by the Emerging Technologies

Hazardous Wastes Destroyed by Molten Salt Combustion

PCB's
Chloroform
Perchloroethylene distillation bottoms
Trichloroethane
Tributyl Phosphate
Nitroethane
Monoethanolamine
Diphenylamine HCl
Rubber tire buffings
Para-Arsanilic Acid
Contaminated Ion Exchange Resins (Dowex and Powdex)
High-Sulfur Waste Refinery Sludge
Acrylics Residue
Tannery Wastes
Aluminum Chlorohydrate

Pesticides and Herbicides

Chlordane
Malathion
Weed B Gon
Sevin
DDT Powder
DDT Powder plus Malathion Solution
2,4-D Herbicide-Tar Mixed Waste

Real and Simulated Pesticide Containers

    plastic, rubber, and a blend of these

Feasible Pesticides and Nitrile Herbicides

Pesticides	Nitrile Herbicides
dieldrin	trifluralin
heptachlor	2,4,5-T dichlorobinil
aldrin	MCPA
toluidine	

APPENDIX C (Continued)

HAZARDOUS WASTES DESTROYED BY THE EMERGING TECHNOLOGIES

Hazardous Wastes Destroyed by Molten Salt Combustion

Phosphorous Insecticides (feasible)

diazinon
disulfonton
phorate
parathion

Explosives and Propellants

TNT
glyceryl nitrate
diglyceryl tetranitrate
glycol dinitrate
trimethylolethane trinitrate
diethylene glycol dinitrate
PETN
DPEHN
Tetryl
Cyclonite
HMX
Composition B

Feasible

JP type hydrocarbon fuel
ethyl alcohol
hydrazine and its derivatives

Chemical Warfare Agents Destroyed

GB
GB Spray-dried salts
Distilled Mustard, HD
VX
Lewisite, L
Toxic Gas Identification Sets (Real and Simulated)
  made of pyrex, wood, plastic, tin-plates steel, and agent

APPENDIX C (Continued)

HAZARDOUS WASTES DESTROYED BY THE EMERGING TECHNOLOGIES

Hazardous Wastes Destroyed by Fluidized Bed Incineration

HCl pickling liquor (spent)
Organotin in spent steel slag blasting abrasive
Organic dye slurries
  red dye slurry (1-methylaminoanthraquinone and starch gum)
  yellow dye slurry (dibenzpyrenequinone and benzanthrone)

Chlorinated Hydrocarbons

PVC waste from a chemical plant
PVC mixed with coal
PVC insulation over copper wire
Chlorinated hydrocarbon waste containing 80% chlorine

Munitions (slurry)

TNT
RDX (cyclotrimethylenetrinitramine)
Composition B

Hazardous Wastes Destroyed by UV/ozonation Technology

PCB's
TCDD (2,3,7,8-tetrachlorodibenzo-p-dioxin)
OCDD (octachlorodibenzo-p-dioxin)
Chlorodioxins (other dioxins are feasible)
Hydrazine
Monomethyl hydrazine
Dimethyl hydrazine (unsymmetrical)
Copper process waste stream
Nitrobenzene

Appendix D
Design-of-Experiment Tests
for PCB Destruction in the Ultrox Pilot Plant

Exp. No.	Effl. Flow Rate l/min	Ozone Flow mg/min	Ozone Conc. wt%	UV Lamp Arrangement 1	2	3	Ozone Mass Flow Distribution 1	2	3	pH Feed	Mid.	Effl.	Temp °C Inf.	Effl.	PCBs-μg/l Infl.	Mid.	Effl.
1	1.0	700	2.0	10	9	10	233	233	233	7.7	8.0	8.0	18.5	30.0	18	0	0
2	3.5	700	2.0	10	9	10	233	233	233	8.0	7.8	7.8	20.0	24.0	32	0.5	0.4
3	2.0	600	1.5	5	5	5	200	200	200	7.8	7.7	7.7	17.0	23.0	26	0.7	0.7
4	2.0	500	2.0	5	5	5	167	167	167	7.7	7.7	7.8	19.5	25.0	34	0.4	0.4
5	1.0	400	1.0	10	9	10	133	133	133	7.8	7.7	7.7	19.0	32.0	30	0	0
6	2.5	600	1.5	10	5	0	200	200	200	7.8	7.8	7.7	17.0	22.0	23	0.3	0.3
7	2.0	700	1.8	0	4	10	233	233	233	7.7	7.7	7.7	18.0	22.0	36	2.1	0.8
8	1.7	300	1.2	10	4	10	100	100	100	7.6	7.6	7.6	13.0	22.0	18	0.2	0.1
9	1.2	200	1.6	10	9	0	67	67	67	7.6	7.6	7.6	18.0	25.0	18	0.4	0.2
10	1.5	500	3.0	0	9	10	200	200	100	7.3	7.5	7.5	16.0	22.0	18	1.0	0
11	2.2	700	3.0	5	7	10	200	200	300	7.6	7.6	7.6	17.0	20.0	34	1.6	1.2
12	1.8	450	2.5	10	9	5	200	150	100	7.5	7.6	7.5	14.0	20.0	34	1.3	0.6
13	1.3	550	2.0	5	9	10	200	200	150	7.6	7.4	7.4	13.0	22.0	20	0.1	0.3
14	0.8	350	1.0	5	5	5	125	100	125	7.4	7.4	7.4	15.0	22.0	46	0.6	0.2
15	1.6	250	1.3	5	5	10	75	75	100	7.4	7.4	7.5	13.0	22.0	46	0.3	1.3
16	1.4	160	1.0	4	4	10	100	30	30	7.4	7.4	7.4	12.0	18.0	46	1.4	1.2
17	2.0	350	3.0	10	8	0	150	100	100	7.4	7.4	7.4	14.0	16.0	35	3.0	3.1
18	1.5	300	2.8	10	0	10	100	100	100	7.4	7.4	7.4	14.0	20.0	35	4.1	2.1
19	1.0	300	2.5	9	4	5	160	80	60	7.4	7.4	7.4	13.0	21.0	26	1.8	1.4
20	1.2	200	1.8	4	4	4	100	50	50	7.6	7.3	7.3	14.0	22.0	26	2.8	1.5
21	0.8	300	2.6	10	0	0	200	50	50	7.5	7.3	7.3	16.0	23.0	23	1.0	1.2
22	1.0	500	3.0	0	0	0	244	122	122	7.3	7.4	7.3	18.0	18.0	23	5.5	4.2
23	1.3	300	1.7	5	0	5	125	50	125	6.0	5.7	5.7	8.0	15.0	42	12.1	2.9
24	1.6	250	1.5	3	3	4	75	75	100	5.7	5.8	5.6	9.0	12.0	31	1.0	0.4

Source: Arisman and Musick, General Electric Company Hudson Falls, New York, Zeff and Crase, Westgate Research Corporation, West Los Angeles, California

APPENDIX D (Continued)

DESIGN-OF-EXPERIMENT TESTS FOR PCB DESTRUCTION IN THE ULTROX PILOT PLANT

Exp. No.	Effl. Flow Rate l/min	Ozone Flow mg/min	Ozone Conc. wt%	UV Lamp Arrangement			Ozone Mass Flow Distribution			pH			Temp °C		PCBs-μg/l		
				1	2	3	1	2	3	Feed	Mid.	Effl.	Inf.	Effl.	Infl.	Mid.	Effl
25	2.0	500	2.6	6	10	2	200	200	100	5.7	5.6	5.6	11	12	42	8.7	2.7
26	1.2	250	2.5	12	30	1	150	150	50	5.7	5.7	5.8	10	18	31	3.7	2.3
27	1.0	400	1.0	29	30	1	133	133	133	6.2	6.2	6.1	15	23	22	2.3	1.2
28	1.5	500	3.0	19	20	1	200	200	100	5.6	5.9	5.9	13	19	22	2.3	1.0
29	1.6	250	1.3	20	30	1	75	75	100	6.3	6.3	6.2	3	6	23	1.2	1.3
30	0.9	300	2.6	10	10	1	50	50	200	7.0	6.8	6.6	14	19	28	3.5	1.1
31	0.6	1000	3.0	29	30	1	300	300	400	7.0	6.9	6.9	14	27	13	1.3	0.4
32	0.6	1000	3.0	29	30	1	400	200	200	6.8	6.9	7.0	10	24	10	0	0
33	0.5	800	3.0	29	30	1	400	200	200	7.0	7.3	7.4	12	28	7	0	0
34	1.4	250	1.3	20	30	1	75	75	100	7.4	7.2	7.2	14	19	18	0.6	0
35	1.6	500	3.0	19	20	1	200	200	100	7.4	7.5	7.5	12	20	18	0	0
36	1.8	250	1.5	10	30	`2	75	75	100	7.4	7.4	7.4	11	16	18	1.0	0.2
37	0.8	350	1.0	14	30	2	125	100	125	7.3	7.2	7.2	15	18	18	0.5	0.4

Source: Arisman and Musick, General Electric Company, Hudson Falls, New York, Zeff and Crase, Westgate
Research Corporation, West Los Angeles, California

Appendix E
Predicted Conditions to Achieve Minimum Costs—Ultrox Plant

Pilot Plant Operating Conditions

Design Number	Effl. Flow Rate 1 ppm	Total Ozone Mass Flow mg/min	Ozone Mass Flow Distribution			Ozone wt%	Number Lamps Total	Number of Lamps Sections			Effl. DEHP g/l	Effl. PCB g/l
			Sect 1 mg/min	Sect 2 mg/min	Sect 3 mg/min			1	2	3		
20006	2.0	200	80	80	40	1.0	15	0	10	5	79	0.7
18430	2.0	100	33	33	33	1.0	15	0	10	5	85	0.9
18254	2.0	100	50	38	12	1.0	15	0	10	5	83	0.9
18222	2.0	100	40	30	30	1.0	15	0	10	5	86	0.9
18414	2.0	100	57	29	14	1.0	15	0	10	5	85	1.0

Source: Arisman and Musick, General Electric Co., Hudson Falls, New York, Zeff and Crase, Westgate Research Corporation, West Los Angeles, California

146